SWATARA
NORTH LEBANON
WEST
LEBANON
ANNVILLE
NORTH
CORNWALL
SOUTH ANNVILLE
SOUTH LEBANON
WEST
CORNWALL
SOUTH LONDONDERRY
LANCASTER COUNTY
20
15
10
MOUNT JOY
PENN
RAPHO
EAST DONEGAL
EAST HEMPFIELD
WEST HEMPFIELD
SUSQUEHANNA
MANOR
LOWER WINDSOR
RIVER
CHANCEFORD
AREA
11 / 79
4 MILES
SCALE
0
2
4
LEGEN
MILES FROM TMI
COUNTY BOUNDARY
prepared for OFFICE OF POLICY AND PLANNING

THREE MILE ISLAND SOURCEBOOK

GARLAND REFERENCE LIBRARY
OF SOCIAL SCIENCE
(VOL. 144)

THREE MILE ISLAND SOURCEBOOK
Annotations of a Disaster

Philip Starr
William Pearman

GARLAND PUBLISHING, INC. • NEW YORK & LONDON
1983

Library of Congress Cataloging in Publication Data
Main entry under title:

Three Mile Island sourcebook.

(Garland reference library of social science; v. 144)
Includes index.
1. Atomic power-plants--Pennsylvania--Accidents.
2. Three Mile Island Nuclear Power Plant (Pa.)
I. Starr, Philip, 1935- . II. Pearman, William A., 1940- . III. Series.
TK1345.H37T48 1983 363.1'79 82-49180
ISBN 0-8240-9184-1

Cover design by Laurence Walczak

Printed on acid-free, 250-year-life paper
Manufactured in the United States of America

To our families, whose support and encouragement made this book possible

CONTENTS

INTRODUCTION

On March 28, 1979, the worst commercial nuclear power plant accident in history occurred at TMI. Although there is still a lack of agreement as to whether a "crisis" was precipitated as a result of the accident, there is no question that the chronological events following the accident have had a significant impact on the nuclear power industry.

The book consists of three distinct sections. First, there is a chronology of selected news media coverage of TMI from the time plans were announced on 11/19/66 to build the plant to the second anniversary of the accident on 3/29/81. The five publications annotated are the *Harrisburg Patriot*, the *Lancaster Intelligencer Journal*, the *Middletown Press and Journal*, the *New York Times*, and *Newsweek*.

The Middletown paper was selected because Middletown is the town where TMI is located even though it is a weekly rather than a daily paper. The *Harrisburg Patriot* is the major newspaper for the capitol city of the state and the largest city within a five-mile radius of TMI. The *Lancaster Intelligencer Journal* represents a near-by large city whose water system is derived from the Susquehanna River, on which TMI is located. The *New York Times* and *Newsweek* represent a national viewpoint.

The second section is comprised of annotations of federal and state government documents that pertain to TMI. Included are the publications of the NRC, Kemeny Commission, congressional hearings, and reports of various departments of the government of Pennsylvania.

The third section is annotations of articles, professional papers, and of books that have been written about TMI. A directory of pro- and anti-nuclear groups and a glossary of terms related to nuclear power generation follow.

The book is intended as a sourcebook for both advocates of or those against the generation of nuclear power. People interested in how events develop over time as well as how the news media covers a significant story will also find the book worthwhile. Finally, the media's coverage of TMI for almost 15 years in five publications can be read as an interesting chronicle of an important historical occurrence.

The authors have tried to keep their private opinions from intruding in either the selection of materials covered or the preparation of the annotations. We wanted to present the facts in an organized, chronological fashion so that readers could decide for themselves what they want to believe about TMI. We are reminded of Sir Julian Huxley's introduction to Father Chardin's book, *The Phenomenon of Man*. Both Huxley and Chardin agreed that life always existed. However, Huxley used this hypothesis to prove that God did not exist, whereas Father Chardin used it to prove the existence of the Lord.

This book would not be possible without the understanding of our home institutions, the H.K. Cooper Clinic of Lancaster, Pennsylvania, and Millersville State College, Millersville, Pennsylvania. Many excellent librarians aided us in our search for resource materials. Libraries utilized in the course of our work include: the Helen Ganser Library of Millersville State College, the Library of the Lancaster Newspapers, the Lancaster County Library, the Library of the *Harrisburg Patriot*, and the Middletown Public Library.

We are also indebted to our secretaries, Anna Brown, Barbara Nehr, and Cristina Traina, who have helped with the various drafts of the manuscript.

ABBREVIATIONS

The abbreviations listed below are presented in alphabetical order to facilitate the reader's comprehension.

AAUW	American Association of University Women
ABCC	Atomic Bomb Casualty Commission
ACRS	Advisory Committee on Reactor Safeguards of the Nuclear Regulatory Commission
ACS	American Chemical Society
AEC	Atomic Energy Commission
AFL-CIO	American Federation of Labor; Congress of Industrial Organization
AIF	Atomic Industrial Forum
ANGRY	Anti-Nuclear Group Representing York
AP	Associated Press
ASLAB	Atomic Safety and Licensing Appeals Board
ASLB	Atomic Safety and Licensing Board
BER	Bureau of Epidemiological Research
BPC	Bechtel Power Company
BRP	Bureau of Radiological Protection
B&W	Babcock & Wilcox
CAT	Capitol Area Transit
CBS	Columbia Broadcasting System
CC	Chamber of Commerce
CD	Civil Defense
CDC	Center for Disease Control

Cong.	Congressman
CIS	Congressional Information Service
CSE	Citizens for a Safe Environment
DC	District of Columbia
DER	Department of Environmental Resources
DOA	Department of Agriculture
DOE	Department of Energy
DPW	Department of Public Welfare
EAF	Environmental Action Foundation
ECNP	Environmental Coalition on Nuclear Power
EDA	Economic Development Administration
EIS	Environmental Impact Statement
EISy	Environmental Information Study
EMA	Emergency Management Agency
EPA	Environmental Protection Agency
ESRP	Environmental Standards Review Plans
Evac.	Evacuation
FBI	Federal Bureau of Investigation
FDA	Federal Drug Administration
FEMA	Federal Emergency Management Agency
FERC	Federal Energy Regulatory Commission
FES	Federal Environment Statement
F&M	Franklin and Marshall College
FPC	Federal Power Commission
GAO	Government Accounting Office
Gov	Governor
GPU	General Public Utilities
HAAANP	Hershey Area Alliance Against Nuclear Power
HACC	Harrisburg Area Community College
HEC	Hershey Electric Company
HEW	Health, Education, and Welfare
HHA	Harrisburg Housing Authority
HMC	Hershey Medical Center

HP	*Harrisburg Patriot*
HRPD	Health Resources Planning and Development
IBEW	International Brotherhood of Electrical Workers
IRS	Internal Revenue Service
JCPL	Jersey Central Power and Light Company
Kemeny Commission	President's Commission on the Accident at TMI
KW	Kilowatt hours
LACI	Lancaster Association of Commerce and Industry
LASL	Los Alamos Scientific Laboratory
LEAF	Lancaster Environmental Action Federation
LEMA	Lancaster Emergency Management Agency
LER	Licensee Event Reports
LIJ	*Lancaster Intelligencer Journal*
LOCA	Loss-of-Coolant Accident
Lt.	Lieutenant
LVC	Lebanon Valley College
LWR	Light Water Reactor
LWV	League of Women Voters
MAA	Middletown Area Association
Mass.	Massachusetts
MCNEC	Middletown Community Nuclear Educational Council
Md.	Maryland
Met Ed	Metropolitan Edison
MIT	Massachusetts Institute of Technology
MPJ	*Middletown Press & Journal*
MSC	Millersville State College
N	Nuclear
NASA	National Aeronautical and Space Administration
NC	North Carolina
NCCANE	New Cumberland Citizens Against Nuclear Energy

NEIL	Nuclear Electric Insurance Limited
NEW	Nuclear Energy Women
NIH	National Institute of Health
NIMH	National Institute of Mental Health
NIOSH	National Institute of Occupational Safety and Health
NJ	New Jersey
NRC	Nuclear Regulatory Commission
NSOC	Nuclear Safety Oversight Committee
NW	*Newsweek*
NYC	New York City
NYSE	New York Stock Exchange
NYT	*New York Times*
ONRR	Office of Nuclear Reactor Regulations, NRC
OPEC	Organization of Petroleum Exporting Countries
Pa.	Pennsylvania
PANE	People Against Nuclear Energy
PEC	Pennsylvania Electric Company
PEMA	Pennsylvania Emergency Management Agency
PHD	Pennsylvania Health Department
PHEC	Philadelphia Electric Company
PILC	Public Interest Law Center
PIRC	Public Interest Research Center
PIRG	Public Interest Research Group
PORV	Pilot Operated Relief Valve
PP&L	Pennsylvania Power and Light Company
PR	Public Relations
Pres.	President
PSS	Pennsylvania Sociological Society
PSU	Pennsylvania State University
PUC	Public Utilities Commission
Rep.	Representative
RLI	Reconnaissance Level Information
SBA	Small Business Administration

SC	South Carolina
SEC	Securities Exchange Commission
Sen.	Senator
SRC	Social Research Center, Elizabethtown College
SRP	Standard Review Plan
SVA	Susquehanna Valley Alliance
Tenn.	Tennessee
TMI	Three Mile Island
TMIA	Three Mile Island Alert
TSAR	Technical Staff Analysis Report
TV	Television
UCS	Union of Concerned Scientists
UPI	United Press International
US	United States
Va.	Virginia
V.P.	Vice-President
VEPCO	Virginia Electric and Power Company
Wash.	Washington

THREE MILE ISLAND SOURCEBOOK

DETAILED CHRONOLOGY OF EMERGENCY RESPONSE

From the time the accident was initiated on March 28, 1979 at 4:00 AM to the lifting of Gov. Thornburgh's Precautionary Advisory on April 9, 1979, the eyes and ears of the world were focused on TMI. For the reader to understand why the accident is considered a "turning point" in the history of nuclear power, a detailed chronology of the highlights of the accident at TMI are summarized.

3/28/79

4:00 AM Accident initiated.

6:55 AM Site emergency declared.

7:24 AM General emergency declared.

7:45 AM PEMA notifies Gov. Thornburgh of plant status.

8:40 AM First team of five inspectors leaves NRC Region 1 for the site.

9:00 AM Pres. Carter informed by Commissioner Victor Gilinsky of NRC.

9:02 AM EPA notified of accident by NRC.

9:02 AM AP releases national bulletin advising a General Emergency has been declared at the plant but that no radiation had been released.

10:30 AM NRC press release confirms incident; release of primary water to containment; no offsite radioactivity declared.

1:15 PM John Herbein, V.P. of Met. Ed., reports at press conference "no significant levels of radiation released; reactor being cooled in accordance with design; no danger of meltdown."

4:30 PM Lt. Gov. Scranton's News Conference-situation more complex than state led to believe-no danger to public health. Met. Ed. has given misleading information-radiation has been released.

10:00 PM Lt. Gov. Scranton's News Conference-high radiation levels on site; no critical levels off site.

3/29/79

5:15 PM Gov. Thornburgh's Press Conference-no cause for alarm; no danger to public health; no reason to disrupt daily routines; situation appears under control, but important to remain alert.

10:00 PM James Higgins, NRC staff, informs Paul Critchlow, Gov. Thornburgh's press secretary, of possibility of extensive fuel damage.

3/30/79

8:01 AM Helicopter measures dose rate of 1200 mR/h over Unit 2 auxillary building vent stack.

8:42 AM BRP and Gov. Thornburgh informed by plant of radiation release.

9:00 AM Wire service story about radiation release.

9:55 AM NRC Commissioners decide evacuation unnecessary.

9:59 AM Gov. Thornburgh informed by Commissioner Hendrie, Chairman of NRC, that evacuation is unnecessary, but advisable for persons within 5 miles of plant to stay indoors.

10:25 AM Gov. Thornburgh on WHP radio advising people within 10 miles of the plant to stay indoors with doors and windows closed.

10:47 AM Pres. Carter orders H. Denton to the site. Denton is a member of the NRC's Emergency Management Team.

12:30 PM Gov. Thornburgh News Conference- no reason to panic, but its advisable for pregnant women and preschool children to evacuate within 5 miles of TMI.

4:00 PM UPI wire story quoting Commissioner Dudley Thompson of the NRC as saying there exists a possibility of core melt-down within a few days.

10:00 PM Gov. Thornburgh and Denton Joint News Conference- The advisory is still in effect. No explosion in the reactor vessel and the possibility of a core meltdown is very remote.

3/31/79

12 Noon Denton Press Conference- Crisis not over; NRC still examining bubble data; does not believe bubble poses a problem.

2:45 PM Hendrie Press Conference- The reactor is in stable configuration and the fuel is cooling down. The possibility of a precautionary evacuation is being explored while the hydrogen problem is being handled.

5:00 PM	Gov. Thornburgh's News Conference- The advisory is still in effect. No necessity of full evacuation; no threat to public health.
11:00 PM	Gov. Thornburgh and Denton Joint News Conference- No possibility of a hydrogen explosion in the reactor vessel in the near future; Pres. Carter will visit 4/1/79.
4/1/79	
1:00- 2:30 PM	Pres. Carter visits the site.
2:00 PM	Denton Press Conference- Action is still underway to eliminate bubble; evacuation still considered unnecessary.
7:00 PM	Gov. Thornburgh News Conference- The advisory is still in effect; government offices will open as usual on 4/2/79.
4/2/79	
11:15 AM	Governor's News Conference- The size of the bubble is decreasing.
4/3/79	
2:40 PM	Denton News Conference- The situation at the plant is stable. Hydrogen explosion is no longer considered a problem.
4/4/79- 4/6/79	Denton News Conference- Only minor problems remain; cooling continues.
4/9/79	Governor's Press Conference- Advisory is lifted.
4/17/79	Denton leaves.

4/27/79 Plant placed on natural circulation cooling.

Source: *Three Mile Island: A Report to the Commissioners and to the Public*, Vol. 2, Part III, Mitchell Rogovin, Director of the Nuclear Regulatory Commission's Special Inquiry Group, 1980.

NEWS MEDIA COVERAGE

Plans for the construction of TMI were announced on 11/18/66. The plant became operational on 9/5/74. Four and one half years later, the worst commercial nuclear accident occurred at TMI on 3/28/79. At the time of the second anniversary of the accident (3/29/81), the plant was still out of commission and the debate over the use of nuclear power for electric generation continued to rage.

In this section, the annotations of news coverage of TMI from 11/18/66 to 3/29/81 are presented. Using information on one nuclear power plant, this coverage will trace the history of nuclear power as the source for cheap, safe electric power to one of controversy as to its safeness, economic feasibility, and the adequacy of emergency planning.

Reporting by the five news media organizations (HP, LIJ, MPJ, NW, and NYT) of the nuclear power question varied over the fourteen plus years covered in this book. Until the accident occurred on 3/28/79, the news coverage was limited. During the accident, there was an explosion of coverage. After the crisis, coverage lessened. This illustrates the crisis orientation of the news media to major issues.

These annotations also raise questions about the ability of governmental institutions to cope with a potential crisis of the magnitude implied in TMI. First, there was unfounded reassurance by authorities that everything was under control. Second, they raise serious questions concerning the ability of the government to safely evacuate the area if needed. Finally, the lack of coordination among government agencies led to contradicting and conflicting reports.

In reading these annotations, the complexity involved in understanding nuclear power is clearly underscored. It leaves the reader wondering who to believe. In the last analysis, the decision to accept the pro or con arguments for nuclear power is based on the assessment of integrity and competence of the presenters. It is not, in a large part, based on an analysis of the issues *per se*.

Since the accident at TMI can be viewed as a "turning point" in the viability of the nuclear power industry, the chronological news coverage of the plant is presented below.

11/19/66 "Met Ed Announces Plan to Build Nuclear Generating Plant."
8,400,000 KW capacity nuclear generating plant at the cost of $100 million on the Susquehanna River at TMI is announced by Met Ed. (LIJ and NYT)*

1/22/67 "Atom Power Plant Planned for Island in PA."
Met Ed plans $100 million plant on TMI on the Susquehanna River. (NYT)

2/13/67 Editorial: "Power on the River."
The editorial emphasizes that the phrase 'the age of nuclear power' is becoming a reality. (LIJ)

2/16/67 "Met Ed to Build Big Nuclear Plant."
W. R. Snyder, Met Ed President, announces that a nuclear power plant will be built at TMI. Plant scheduled for completion in 1971. (MPJ)

* Entries with and/or also indicate that a story similar to the one annotated appears in the other paper(s).

3/9/67 "Island Cottages Must be Removed by Nov. 1st."
Owners of about 60 cottages located on TMI were notified that their cottages must be removed by 11/1/67. (MPJ)

"Nuclear Power Plant at Three Mile Island Viewed as Boon to Area."
Met Ed officials explained TMI plans at a special Borough Council meeting. They envision the TMI plant as an economic boon for the area. Details were provided on procedures that must be complied with for the nuclear plant to be constructed. Construction to begin in early 1968. (MPJ)

3/24/67 "'Hot Strip'-6 Million KW of River Power."
A boiling water reactor at a projected cost of $350 million is being planned and will be constructed so that the TMI plant can't behave like a bomb. Power officials also state that the need for electrical power is far exceeding present generating capacities. (LIJ)

4/27/67 "Training, Exacting Research, Tons of Reports Face Utility in Building Atomic Electric Plant."
Reporter's conversation with TMI vice-president in charge of engineering relative to nuclear safety and planning of the TMI facility. Several aspects of GPU involvement in nuclear power development are reported. Some statistics on atomic plants in the U.S. are presented. (MPJ)

5/9/67 "Met Ed Applies for Power Plant Permit."
Met Ed sends an application for permission to construct the TMI plant to the AEC. (LIJ; also 5/11/67)

7/22/67 "Construction at Three Mile Island to Begin."
Work to begin next week. Expected completion date is 12/1/71. (LIJ)

7/27/67 "Met Ed Subsidizes Archeological Project at Nuclear Plant Site."
Archeologists from the Pa. Historical and Museum Commission are working against time seeking to recover evidence of the Indian Culture on TMI. (MPJ)

9/26/67 "Work Progressing on Temporary Span to Three Mile Island."
A span to move heavy equipment to TMI across the Susquehanna River is progressing. The span is temporary. (LIJ; also 9/28/67 MPJ)

11/23/67 "Chamber Endorses Truck Legislation: Hear Report on Met Ed Nuclear Plant."
Met Ed presented a report on TMI history and future plans to the Steering Committee of the MAA of the Harrisburg Area CC. (MPJ)

2/16/68 "Public Hearing on Power Plant is Proposed."
The AEC has scheduled a public hearing in Middletown for 4/10. The hearing will consider Met Ed's application to build a nuclear power plant at TMI. (LIJ; also 2/22/68 MPJ)

4/11/68 "Federal Agency Holds Public Hearing on Utility Proposal: Bid to Build Nuclear Facility Near Middletown Unopposed."
Reports on AEC hearings on Met Ed's bid to build a nuclear power plant. No public opposition voiced. Evidence in favor of the plant is presented by Met Ed and the AEC staff. The issue of aircraft collision is raised. (HP)

5/16/68 "The AEC Grants a Provisional Construction Permit to Met Ed."
A provisional construction permit for the construction of a nuclear power plant on TMI is granted. Estimated cost will be $110 million. The staff of the AEC stated that the "proposed facility can be constructed at the proposed site without undue risk to the health and safety of the public." (LIJ)

9/16/68 "Work Underway on Three Mile Island Nuclear Plant."
Construction on Met Ed's $136 million nuclear generating plant has started. It is expected to be completed in late 1971. (LIJ)

10/16/68 "Nuclear Generating Station Grows on Three Mile Island."
The article reports on the progress of construction at TMI. (MPJ)

1/8/69 "Met Ed Will Build 2nd Nuclear Plant at Three Mile Island."
A second nuclear unit originally scheduled for Oyster Creek will be built at TMI. More than 1.1 million KW of generating capacity will be added to GPU's system. (MPJ)

3/5/69 "Employment at Nuclear Power Station Climbs Above 700."
Met Ed has budgeted $48.2 million for the 1969 portion of TMI construction. Over 700 people now work at TMI and several hundred more are expected by the end of summer. (MPJ)

"Model of Three Mile Island Nuclear Plant Shown."
A model of Met Ed's TMI plans is on display at the Middletown Borough Office. (MPJ)

4/23/69 "Lions to Hear Talk on Atomic Power Plant."
The construction project manager at TMI will address the Middletown Lions Club. (MPJ)

6/14/69 "Strike at Three Mile Island Nuclear Plant is Settled."
A two day strike at Met Ed's TMI construction site is settled. The dispute was over jurisdiction. The plant is 15% completed. (LIJ)

6/18/69 "1,024 Working at Nuclear Plant-Payroll is $250,000 a Week."
The article describes Met Ed's construction as part of an area-wide economic boon. It also gives details of the construction to date. Details of future construction are provided. (MPJ)

7/29/69 "Advisory Committee on Reactor Safeguards Approved Plans to Construct a Second Nuclear Power Plant at TMI."
The ACRS recommends that the plant will not be an "undue risk to the health and safety of the public." The cost will be $193 million and it is scheduled to be finished in 1973. (LIJ)

8/29/69 "New Atomic Plant Rising Near Bainbridge."
The photo shows progress in construction. The story announces the AEC's hearing on Unit II at TMI will be held on 10/6 in Middletown. (MPJ)

10/8/69 "N-Power Plant 'No Hazard to Low Flying Aircraft.'"
A state geologist fears that TMI cooling towers could be a hazard for low flying planes. He also fears vapors emitted from towers would add to aircraft landing problems. Met Ed feels chances of a crash are one in a million per year. Met Ed also states that the plant is designed to withstand tornadoes and that the

damage to one of the four towers would not release any radioactivity. (MPJ)

11/7/69 "Construction Permit Granted to Met Ed." A permit to construct Unit II nuclear generating power plant was granted by the AEC to Met Ed. The cost is estimated at $236 million and it will be completed in 1973. (LIJ)

1/14/70 "Met Ed Observation Center Opens Feb. 8." Plans are announced for the opening of Met Ed's Observation Center overlooking TMI. (MPJ; also 2/2/70 LIJ)

5/6/70 "Engineers Report on Cooling Tower Vapor." Three engineers working on the Met Ed project report that the cooling towers at TMI will have little atmospheric effect at ground level and a limited effect at higher levels near the plant. (MPJ)

5/20/70 "42 Area Men Training to Help Operate 3 Mile Island Plant."
Area men are being trained to become control room operators and auxilliary operators for TMI. Details of pre and operational testing are reported. (MPJ)

5/26/70 "Safety Gets Priority at New Atomic Plant."
The article describes what Met Ed is planning to do to prevent radiation leakage, handling of radiation waste, and the possibility of an airplane crash. It concludes "Met Ed cannot afford a disaster at its TMI plant even if the human element is discounted. There are too many million dollars involved." But, Met Ed says the level of radiation released will be far below AEC mandated safety levels. (LIJ)

7/1/70 "Met Ed Gets $35,000 for River Study." Met Ed is granted $35,000 for a preliminary chemical and biological tests of the Susquehanna River prior to the scheduled operation of the company's nuclear power plant under construction at TMI. Dr. David Kurtzman, Secretary of Education for Pa., signed the agreement. (LIJ)

7/15/70 "Progress at TMI."
Picture and description of the new TMI Observation Center and a view of the plant. (MPJ)

7/29/70 "Strike at Fruehauf: It's Quiet at '3 Mile.'"
Production workers at TMI agreed to return to work after one day job action. The issue was over who should run compressors, engineers or painters. (MPJ)

8/5/70 "Press Representatives Get Tour-Report on Three Mile Island Plant Progress."
A report on a Met Ed sponsored tour of TMI and progress in construction to date. (MPJ)

8/7/70 "Reactor to Operate Nuclear Power Plant at Three Mile Island Arrives."
The reactor was transported without nuclear components. At times, the reactor's weight destroyed the roads it passed over. (LIJ; also 8/16/70 NYT and 8/26/70 MPJ)

9/2/70 "Met Ed Applies for Power Line."
Met Ed has asked Dauphin County for permission to construct a power line for the operation of TMI. (LIJ)

9/9/70 "Nuclear Plant Focal Point for Conference."
Met Ed will conduct a nuclear science conference for high school students and teachers. TMI will be the focal point. (MPJ)

9/30/70 "Work-Week Cut at Three Mile Island."
The work week at TMI has been cut from 50 to 40 hours per week. The company says this is not indicative of a shut-down, but simply due to the forthcoming darkness of days. (MPJ)

10/7/70 "Science Seminar will be Held at Three Mile Island Nuclear Station."
Ninety area teachers will be hosted by Met Ed at TMI in the first phase of a Nuclear Science Conference. (MPJ)

10/14/70 Editorial: "Pollution Fears Said to be Unfounded."
The editorial concludes that the danger of radiation from a nuclear plant is equal to wearing a watch with a radium dial. (MPJ)

1/20/71 "Fire Destroys 2 Structures on 3 Mile Island."
A frame building and a trailer office at TMI were destroyed by fire. (MPJ)

2/4/71 "Met Ed to use River to Cool Nuclear Power Plant."
Met Ed has filed an application with the U.S. Army Corps of Engineers' district office in Baltimore for a permit to discharge 35,000 gallons of water a minute into the Susquehanna River from the new plant. If approved, a Met Ed spokesman stated that the evaporation plume would produce fog that might affect air traffic at the nearby Olmstead Airport for about 70 total hours per year. (LIJ)

2/17/71 "Ol' Man Winter."
A photo story on construction progress at TMI despite the cold winter. (MPJ)

"It's No Mine."
A photo story on the vent being used to circulate fresh air at the TMI complex. (MPJ)

"Large Crane and Shovel."
A photo story on the removal of a dam at TMI. (MPJ)

4/7/71 "Construction."
A photo story showing Unit I 70% complete and Unit II 15% complete. (MPJ)

4/14/71 "Nuclear Science Conference at Lebanon H.S."
Met Ed sponsored a Nuclear Science Conference at Lebanon High School. Dr. W. Witzig of PSU, Chairman of the Nuclear Engineering Department, stated that the nuclear reactor can't become a plutonium bomb as the device shuts itself down if it overheats. Dr. G.H. Whipple, Radiological Health Professor at the University of Michigan School of Public Health, is conducting studies to assess if a nuclear power plant can be safely operated at TMI. Dr. Leonard Sagan, of the Palo Alto Medical Clinic, Environmental Medicine, stated that radiation from x-rays is a greater danger than that from a nuclear powered plant. (LIJ)

5/5/71 "New Addition."
The photo story discusses new towers to carry high voltage lines for TMI. (MPJ)

6/9/71 "Atomic Plant Here Hit by Work Stoppage."
Operating engineers began picketing TMI and Peach Bottom plants; construction is halted. (MPJ)

6/23/71 "3 Mile Island Work Slowed by Strikers."
An engineers work stoppage continues to slow TMI construction. (MPJ)

6/30/71 "Little Activity."
A photo story on the work slowdown at TMI as viewed from the air. (MPJ)

7/14/71 "Work Force Returning to 3 Mile Island."
Operating engineers approve a new contract; work to resume at TMI. (MPJ)

"New Transmission."
A photo story on new power transmission lines at TMI. (MPJ)

7/28/71 "Met Ed's."
A photo story on the draft cooling towers at TMI designed to prevent thermal pollution of the Susquehanna River. (MPJ)

8/11/71 "Susquehanna River Fish Study Underway at Three Mile Island Nuclear Station."
A study aimed at obtaining background information on fish life in the Susquehanna River is underway. (MPJ)

8/25/71 "370 Ton Reactor Vessel Positioned at TMI Nuclear Plant."
A photo story of men and machines positioning a 370 ton reactor vessel at TMI. (MPJ)

9/9/71 "Another Big Lift Coming."
A photo story showing easing the steam generator into the reactor building. (MPJ)

"Court Ruling to AEC not Expected to Delay 3 Mile Island Project."
An AEC order from a federal appeals court to reconsider the environmental impact of 96 atomic power plants is not expected to affect the TMI project. (MPJ)

9/15/71 "No Moon Launch Here."
A photo story on positioning a steam generator at TMI. (MPJ)

12/2/71 "Nuclear Plant Project Held Up."
The AEC holds up construction of Unit II pending completion and review of an EIS. Construction of off-site portions of power transmission lines at Unit II are affected. (LIJ)

1/2/72 "Like Inverted Drinking Cups."
A photo story on the shape and visibility of TMI towers. (MPJ)

1/5/72 "Met Ed Plans Major Recreation Area for Public Use on TMI."
Met Ed announces plans for a major east coast recreation area at TMI. The opening is estimated for 1975. (MPJ)

1/12/72 "Taking Shape."
A photo story on electrical sub-station construction. The station will handle electricity from TMI. (MPJ)

"High Reactor Vessel."
A photo story showing transporting a reactor vessel to TMI from Indiana. (MPJ)

1/26/72 "Preparing for Big Lift."
A photo story on the change in placement of the reactor for Unit II. (MPJ)

2/2/72 "Wirlybird Finesse."
A story of helicopter lifting cranes used in construction of TMI towers. (MPJ)

2/16/72 "Nuclear Power: Safe and Reliable."
GPU spokesman says nuclear power is safe, reliable, and economical. He calls TMI the largest GPU power project to date. The construction schedule is discussed. (MPJ)

"Enlarged Photograph."
The photo story is on an MAA meeting where an enlarged photo of TMI was exhibited. (MPJ)

3/15/72 "Moving 600 Tons."
A photo story on the placement of a steam generator to its permanent position at TMI. (MPJ)

"AEC Approves Line from TMI."
A 500 KW transmission system has been approved for TMI by the AEC. (MPJ)

6/16/72 "Year and Half Lost on New Atomic Plant." Unit I is 80% complete and Unit II is 20% complete. Unit I is expected to operate late in 1973. This represents a one and one half year delay. The estimated cost is $700 million. The preparation and review of an EIS resulted in the delay. The AEC did give permission for construction to go ahead. (LIJ)

6/21/72 "2700 Work at 3 Mile Island; Weekly Payroll at $1 Million." Details of TMI construction schedules and how the project has contributed to the economy of the area. (MPJ)

8/30/72 "Hearings Urged on Nuclear Plant." CSE of Harrisburg and the ECNP of Philadelphia have asked the AEC to hold a public hearing on TMI before granting an operating license. The hazard of placing the facility in a flood-prone and heavily populated area is questioned. The groups also cite other problems such as possible design construction and placement deficiences of radiation containment. Fuel handling and emergency core cooling systems are also noted.

8/31/72 "At Three Mile Island: EPA Asks 'Cleaner' Atomic Discharge." Plans are made for TMI to meet EPA standards for the disposal of radioactive waste. (HP)

9/13/72 "Workmen at 3 Mile Island." A photo story on water intake facility construction at TMI. (MPJ)

9/15/72 "Met Ed Issue is Approved." The PUC authorized Met Ed to issue $53 million in bonds to repay short term loans and to pay part of its construction costs for TMI. (LIJ)

9/27/72 "Environmentalists Question Safety of 3 Mile Island Nuclear Power Plant." Belated furor over TMI safety and ecological impact is reported. Newspaper and TV campaigns against the plant are cited. (MPJ)

12/13/72 "TMI Project Gets Green Light." The AEC issues a final EIS on construction and operation of Units I and II at TMI. The AEC recommends that construction continue. (HP)

2/27/73 "AEC Approves: Group to Intervene in Plant Hearings." CSE will be permitted to intervene in hearings on granting an operating license for the nuclear power plant on TMI. The hearings concern the licensing of Unit I. (HP; also 2/28/73 LIJ)

3/13/73 "N-Plant is Outlined for Students." Capitol Area Youth Forum Student Cabinet Members representing 34 senior high schools in Central Pa. met with TMI's Public Information Coordinator to express their concerns about the effects of TMI. (HP)

4/20/73 "Environmental Groups Appeal Reactor Ruling." CSE and ECNP ask a federal appeals court to stay AEC proceedings involving the two reactors at TMI. The groups require technical assistance and expertise but only have $2,000. (HP)

5/23/73 "AEC Schedules Prehearing on TMI Nuclear Plant." The AEC has scheduled a prehearing on the licensing of Unit I. (MPJ)

5/25/73 "At Three Mile Island: Group Asks Halt on N-Plant Work." CSE wants a prehearing conference prior to the actual public hearings on the licensing of TMI. The group had submitted 73 contentions of the plant being unsafe. CSE, along with ECNP, filed suit in federal court seeking funds and technical assistance. They claim that they are entitled to same under the National Environmental Policy Act. (HP)

5/31/73 "Suit Delays TMI Plant."
A court suit by two environmental groups will cause a long delay in a hearing for an application to operate a nuclear power plant on TMI. The groups are asking the AEC to award them $80,000 to pursue in detail the 70 complaints they have against the power plant. Of the 70 complaints, 61 were dismissed at a prehearing as being repetitious. The remaining complaints are as follows: 1) Reaction cooling systems are not designed and constructed so that they will function in an emergency; 2) AEC rules and regulations for radiation releases are improper and contrary to the health, welfare, and safety of the public; 3) EIS fails to discuss the effects associated with transportation, storage, processing, and monitoring of nuclear fuel. (LIJ)

6/29/73 "MSC Gets Grant to Study Power."
MSC has been awarded a grant of $70,507 from Met Ed to conduct the fourth year of a chemical and biological study of the Susquehanna River at the nuclear power plant at TMI. Dr. Donald Davis, MSC Chemistry Professor, is the coordinator for the project. (LIJ)

7/12/73 "Power Firm Proposes $750,000 Recreational Area."
The article details power company plans for recreational development at TMI; several years in the future. (HP)

7/18/73 "Cooling Tower Emissions Won't Affect Jetport."
The article details why vapor plumes from TMI towers will not affect the area or nearby jetport. (MPJ)

"AEC to Hear Reports on TMI Operating License."
The process for testifying before the AEC on TMI Unit I operating license request is discussed. (MPJ)

"Nuclear Briefing."
A briefing on environmental safeguards was held for the Interstate Legislative Committee on Lake Erie at TMI. (MPJ)

8/28/73 "Utilities' Lawyers Ask Expeditious Handling: Three Mile Island Hearing Date Sought."
Power companies request that the AEC move up the date of hearings on licensing. A prehearing conference was held to lay groundwork for a public hearing on Unit I. (HP)

8/30/73 "Power Plant Hearings Eyed."
The AEC Regulatory Staff recommended hearings be set for November on the operation of a reactor at TMI. The ASLB will hold the hearings and make the decision. (LIJ)

9/12/73 "AEC to Hold Hearing on Power Line for TMI."
The AEC will hold a hearing in connection with transmission line construction for Unit II. (MPJ)

10/31/73 "AEC to Hold Hearing on TMI."
AEC hearings on a license to operate Unit I will begin 11/5 in Harrisburg. (MPJ)

11/6/73 "Nuclear Plant Safety Panel Delays Decision on Hearing."
Owners of TMI have requested that Pa. Insurance Commissioner Herb Denenberg be barred from testifying at licensing hearings. Charles A. Hasking, Licensing Board Chairperson, delayed a ruling on the request. (HP)

11/7/73 "N-Plants Called 'Doomsday' Devices."
The article reports on statements made by "intervenors" at hearings of the ASLB relative to licensing of the TMI nuclear power plant. (HP and MPJ).

"TMI Nuclear Unit Safety Hearings Underway."
A report on the opening day of hearings by the AEC for the licensing of Unit I. (MPJ)

11/8/73 "N-Plant Opposition Withdrawn."
In the third day of hearings, relative to licensing TMI, opposition was withdrawn by the intervenors (Harrisburg CSE and the ECNP of Philadelphia) in exchange for the installation of a "charcoal kidney filter" to reduce low-level radiation. The article contains comments by Herb Denenberg against granting a license. (HP and LIJ; also 11/28/73 MPJ)

12/8/73 Editorial: "Learning to Compromise."
Praise for compromise is given because an unnecessary delay was avoided. The safety device will be installed and the environmental groups still retain the "right to seek revocation, suspension, or modification of the operating license after commencement of operation." (LIJ)

12/12/73 "TMI's First Unit 98% Complete."
Unit I is 98% complete. Unit II is 37% complete. There is also a report on the investment in company and mortgage bonds. (MPJ)

12/29/73 "Denenberg Cites AEC Safety Complaints: Further Hearings on N-Plant Asked." Denenberg seeks to reopen public hearings on TMI construction; cites AEC criticisms of the nuclear power industry. Changes in safety rules issued by the AEC are listed. (HP)

1/4/74 "New Superintendent at TMI." John Herbein will replace Richard M. Klingman as Superintendent of TMI. (MPJ)

1/9/74 "First Shipment of Nuclear Fuel for Reactor at TMI Due January 10." The first shipment of nuclear fuel will arrive at TMI. Fuel assemblies are being shipped from B&W in Lynchburg, Va. The chain reaction process within the reactor is explained. (MPJ)

2/13/74 "Fire." The photo story shows a fire at TMI. A stove ignited propane containers. Following the blaze in the storage trailer, Middletown Borough and Londonderry Township officials will meet with Met Ed representatives to define a policy of fire protection for the nuclear site. (MPJ)

2/20/74 "AEC Testing Safety Systems at Island." Plant safety systems at TMI are in the process of being tested by the AEC. Also, there is a discussion of emergency planning and evacuation plans. (MPJ)

4/3/74 "Unit No. 1 at TMI Nears Completion." Unit I is reported as 98% complete. Unit II is 43% complete. (MPJ)

4/10/74 "TMI." A photo showing a panorama shot of TMI. (MPJ)

4/17/74 "All Unit No. 1 Fuel at TMI; Met Ed will Check Air Quality."
The last nuclear fuel assembly has arrived for Unit I. The shipment and radioactivity checks are discussed. A radiation emergency drill was held. (MPJ)

4/20/74 "Output Limited: TMI Issued Go-Ahead."
A full power operating license for Unit I was issued by the AEC. The plant is restricted to 90% of rated power output until repairs are made to a damaged containment spray tank. The license is for 40 years, beginning 5/18/68. (HP)

5/24/74 "Met Ed N-Facility Mishap Halts Start."
An electrical fire will delay the start-up of Unit I. (HP and LIJ; also 5/30/74 MPJ)

6/1/74 "AEC Plans for Review of N-Plant."
The AEC will review Met Ed plans to operate Unit II at TMI. The public can file a petition to intervene. (HP)

6/7/74 "Paves Way for Power: Nuclear Reaction Occurs at Island."
A chain reaction generating heat necessary to convert water to steam was produced in Unit I reactor. (HP)

6/12/74 "Nuclear Reaction Triggered at Met Ed's TMI."
At 10:36 PM on 6/5/74, the first self-sustained reaction was achieved in Unit I. (MPJ)

7/31/74 "Steam Release Creates Noise at TMI."
Actuation of steam release valves creates loud noises in the TMI area. (MPJ)

8/7/74 "AEC Plans Hearing for TMI's Unit 2."
Environmental matters will be the subject of public hearings on Unit II by the AEC. (MPJ)

9/5/74 "TMI Operating 'Commercially.'"
At 12:01 on Labor Day, Unit I went on line. (MPJ)

9/12/74 "Symbols of Power."
A photo story on white vapor coming from TMI cooling towers. (MPJ)

9/14/74 "22 Violations Reported; TMI Citations Confirmed."
TMI is issued 22 citations for safety violations in 14 months. The article describes the violation rating scale used by the AEC. These violations were categorized as 2 and 3. Category 1 is the most serious. (HP and LIJ)

9/28/74 "Capital Outlay Budget Cut by GPU."
GPU announced it plans to cut $394 million from its capital outlay budget for 74-76. The decision will delay the completion of TMI Unit II until the spring of 77. The plant is 60% complete. (NYT)

9/30/74 "Met Ed Fined for Violation."
Met Ed is cited and fined $4,000 for a security violation at TMI. An AEC inspector was able to enter the grounds without being stopped. Met Ed says the deficiency has been corrected. (LIJ)

10/1/74 "Met Ed Construction on River Delayed Year."
Met Ed announced that the completion of Unit II at the TMI nuclear generated power plant is being delayed one year as part of a general austerity program. (LIJ; also 10/2/74 MPJ)

10/16/74 "Radioactive Gas Release No Hazard."
Radioactive releases from a reactor auxiliary building are said not to pose any hazard to the public. (MPJ)

10/23/74 "N-Plant Reporting Plan Considered After Mishaps."
The Pa. BRP is disturbed following a recent release of radioactive gas at TMI. (MPJ)

10/24/74 "N-Plant Shut Down Due to Leak of Gas."
The TMI plant was shutdown to repair a continued, harmless leakage of radioactive gas, according to a Met Ed spokesman. (LIJ; also 10/25/74 HP)

11/20/74 "N-Unit Shutdown at TMI."
Unit I reactor was shutdown when a control rod drive motor failed. Five days of shutdown will be needed to correct the problem. (LIJ and MPJ).

1/12/75 "Plant Link Unsure: 'Hot' Isotypes Found in Mud."
Traces of 4 radioactive isotopes were detected in Susquehanna River sediment. One, Cesium 134, might have come from TMI. Discrepancies in operation concerning valve failure, and difference in measured and calculated "reactor core vertical power distribution" were also reported. (HP)

1/24/75 "N-Facility Shut Down by Met Ed."
Malfunction in a turbine-related system caused shutdown at TMI. (HP)

1/27/75 "N-Plant Gas Leak Not Threat."
A Met Ed spokesman stated that the "radioactive gas leak was due to equipment problems and didn't threaten public safety." (LIJ)

1/28/75 "Radioactive Gases Released at N-Facility, Met Ed Admits."
Met Ed issues a new statement on the release of steam at TMI last week. The statement admits the gas was radioactive. (LIJ)

1/29/75 "TMI Returns to Service After Brief Shutdown."
Unit I's turbine generator was tripped off line due to equipment problems. The article contains some evaluation of Unit I. (MPJ)

1/31/75 "Area Ruled Out as Site for Nuclear Energy Park."
Public utilities serving South Central Pa. decided that nuclear plant concentration is "excessive" and the area should not be a site for energy park development. (LIJ)

3/9/75 "Federal Study Charges Little Concern by Utilities with Reactor Reliability."
In response to the UCS, Dr. Triner of the NRC, finds that utilities that own most American nuclear reactors, are not sufficiently concerned about safety and performance. Most PUC's have little or nothing to say about safety or design. (NYT)

3/12/75 "March Fuel Charge Cut by Met Ed."
The efficiency of TMI's nuclear plant is cited by Mr. W. Crietz, Pres. of Met Ed, as one reason for the drop in fuel charges. (LIJ)

3/29/75 "Met Ed Wants 3 Mile Water Rules Relaxed."
Met Ed is appealing to the Federal EPA. The basis of appeal is the unfairness of the criteria for discharging heated water from TMI into the Susquehanna River. The EPA is concerned about thermal pollution. LEAF has asked to become a party to the appeal. The kind of appeal is still undecided. (LIJ; also 4/2/75 MPJ)

4/8/75 "At TMI; N-Plant Emergency Drill Set."
A radiation emergency drill will be conducted. The drill will monitor the

plant and surrounding area, and carry out detailed emergency monitoring procedures to insure readiness in the event of an actual or potential emergency. The article also announces an out-of-service period for maintenance and refueling. (HP)

4/8/75 "GPU Warning."
GPU Pres. William Kuhns warns that electric power costs will go higher if restrictions on nuclear power plants are implemented. Restrictions being proposed resulted from a fire at Brown's Ferry plant. (NYT)

4/9/75 "Full Service Will Resume at N-Plant."
Electric motor failure, which is not a part of the nuclear reactor system, caused a three day shutdown at TMI. The plant will resume service and the scheduled emergency drill will be held. (HP)

5/14/75 "TMI Produced Over 4 Billion KW Hours."
Since going into commercial operation on 9/24/74, Unit I has produced over 4 billion KW hours of electricity. (MPJ)

"TMI to Shutdown for Changes."
Unit I's service reliability record is 91%. Unit I will be temporarily shutdown to effect changes to the sequence in which control rods are withdrawn from the reactor. (MPJ)

5/29/75 "N-Station Shutdown Continues."
The plant had been scheduled to shutdown for normal maintenance, but a problem with the pump relating to heat dissipation forced an early shutdown. The article also included reports on a radioactive gas release within one building. (HP and LIJ)

6/4/75 "County Parks Board to Visit TMI."
The Dauphin County Parks and Recreation Board will tour TMI facilities and review plans for public recreation areas adjacent to the plant. (MPJ)

6/5/75 "Security Attacked at TMI."
Charges were made by former guards at TMI. Both were fired after reporting security problems to their superiors. Ralph Nader, consumer advocate, joined in calling for an investigation of security. (LIJ)

6/6/75 "Three Mile Plant Attacks Tied to Dispute."
Met Ed spokesperson alleges that disputes between two security companies led to allegations by Ralph Nader about what appeared to be security breaches at the plant. Three former security guards spoke at a press conference with Nader and indicated there was a conspiracy to perform acts that would create an appearance of a security breach. (HP)

6/10/75 "Despite Firms 'Rivalry' at Three Mile Island: Agency Says N-Plant Security Meets Rules."
The NRC knew about disputes between the two security companies that guard TMI, but does not believe the rivalry jeopardized the plant's security. Details of allegations and comments on security dispute are presented. (HP)

6/11/75 "Probe Set of Security at N-Plants."
The State PUC decides to investigate all nuclear power plants in Pennsylvania. Majority member, Robert K. Bloom of the PUC, is skeptical of the value of the investigation. He stated that the NRC should look into the situation. (HP and LIJ)

6/16/75 "3 Mile Reactor in Service."
Unit I resumed service today after being shutdown for pump repair and rod control changes since 5/25. (LIJ)

6/18/75 "On a Clear Day."
A photo story of the view of TMI from the shoreline at Goldsboro. (MPJ)

6/25/75 "Can Terrorist Sabotage Nuclear Plant on River?"
A three hour tour of the facility revealed that the security system at TMI could be overrun by a determined and well-armed group. (LIJ)

7/12/75 "On N-Plant Snags: Met Ed Outlines Reports to U.S."
Details of four incidents at TMI that required reporting to the NRC were outlined in a public report by Met Ed. Incidents involved: 1) run-off and increased suspended solids in the Susquehanna River as a result of heavy rains; 2) a reactor high pressure trip setpoint which was found to be slightly out of specification; 3) a broken connection in one of several duplicate reactor temperature instruments; 4) increase of power level following an outage prior to assuring that all effects of previous operations were stabilized sufficiently. (HP; also 7/14/75 LIJ and 7/16/75 MPJ)

7/19/75 "Clean Bill for Met Ed."
A routine inspection by NRC officials resulted in the assessment that security operations at TMI are adequate to protect the health and safety of the public. (LIJ)

7/28/75 "Is Dispute by Guards Endangering N-Plant?"
Met Ed stated that security is not foolproof but backup systems are excellent to prevent the nuclear reactor from doing harm. (LIJ)

8/2/75 "Power Cut at Met Ed."
Met Ed indicated that the nuclear generating station was reduced by 25% for three days due to a problem with the steam generator feed pump. (LIJ; also 8/6/75 MPJ)

8/12/75 "Three Mile Island Now at 96%."
Met Ed reported that the nuclear generating plant operated at 96.5% of its capacity for July resulting in lower fuel adjustment charges. (LIJ; also 8/13/75 MPJ)

8/20/75 "3 Mile Saves 10 Million Barrels of Oil."
A report on the first 11 months of TMI operation. Electric output estimated to have saved the purchase of 10 million barrels of oil. (MPJ)

8/29/75 "3 Mile Island Security Criticized."
Examination of Met Ed's "docket files" reveals that it has been cited 33 times for safety violations and infractions by the NRC. Fifteen were in the lowest severity category whereas 18 were in the intermediate level (could be dangerous if not corrected). (LIJ)

9/4/75 "Area Nuclear Plant Marks First 'Birthday.'"
TMI completed one year of nuclear generation on Labor Day. (MPJ)

9/17/75 "Good Record Compiled by TMI's Unit I."
Discusses TMI record of electrical generation and how it contributes to overall nuclear production of electrical power in the U.S. (MPJ)

10/8/75 "Storm-Flood did not Affect Three Mile Island."
Tropical storm Eloise has no affect on TMI. (MPJ)

10/14/75 "Nuclear Plant to be Refueled."
Met Ed stated in mid-January TMI will be shutdown for 8 weeks for refueling. One-third of the 177 fuel assemblies in the reactor core of Unit I will be replaced at a cost of $10 million. (LIJ; also 11/5/75 MPJ)

10/15/75 "TMI Plant had Good Month in September."
The availability factor for TMI in September was 86.7%. Overall the plant averaged 87% during the first 13 months. (MPJ)

11/26/75 "First Fuel Arrives for 3 Mile Island."
The first of five shipments of fuel assemblies for the refueling of TMI in January 1976 has arrived. (MPJ)

11/29/75 "At Three Mile Island: Radioactive Gas Leak Reported."
Radioactive material escaped into the air and water. Met Ed says health and safety are not threatened by the escape, which occurred sometime within the last two weeks. (HP and LIJ)

1/7/76 "TMI Station I Operating at 100% Power."
Since December 24, TMI Unit I has operated at 100% of power. (LIJ)

1/15/76 "Nuclear Evacuation Plan Announced."
Lancaster County CD has developed a 44 page emergency evacuation plan to be used if a disaster occurs at TMI nuclear generating station. Paul Leese is director of CD and Dr. William Conners is his deputy director. (LIJ)

1/21/76 "On a Clear Day You Can See Three Mile Island."
The photo story is on a highway project featuring the TMI skyline in the distance. (LIJ)

2/2/76 "Security is Breached at 3 Mile Island." Security was breached at TMI when an unauthorized construction worker illegally gained access to the plant complex and spent more than an hour, alone, in restricted areas of the station. The construction worker was fired. (LIJ)

2/3/76 Editorial: "Too Easy to Breach Nuclear Plant Security."
A review of past security breaches at TMI led LIJ editors to conclude: "Will we get a full report; one that will assure us that the NRC will order measures at TMI designed to better protect us. We'll be waiting." (LIJ)

2/16/76 "N-Plant Break-In Reviewed."
Met Ed sent a report of the past security breach to the NRC. It concluded that the violation "did not relate to the security of the island." Met Ed indicated that additional training of workers was needed. (LIJ)

2/17/76 Editorial: "Nuclear Plant Security."
Met Ed is engaged in double-talk concerning the security violation. Met Ed's statement regarding the violation "did not relate to the security of the situation." It indicates that the company is "insensitive to the public's deep concern about this matter, and that this insensitivity feeds the understandable fear people have of nuclear generating facilities." (LIJ)

2/20/76 "Jan. 27 Incident at Three Mile Island Probed: Intruder Eluded Guards in N-Plant."
The details of an account of an intruder entering TMI and eluding security guards.

2/20/76 "N-Plant to Close 7 Weeks to Refuel."
Met Ed will shutdown the plant at TMI on Saturday for 7 weeks for refueling. (LIJ)

2/21/76 "2 N-Plants Run Below Capacity."
Met Ed released statistics that the nuclear plant at TMI operated at an overall 77.3% of capacity for 1975. Friends of the Earth, an environmental group, reported that the average capacity operation is about 60%. At this rate, the cost of operating a nuclear powered plant is staggering and "can only be met by large government subsidies to nuclear industry." (LIJ)

3/15/76 "4 Incidents Reported at TMI Plant."
Met Ed reported that "unforeseen and unexpected" problems will extend the refueling process until 4/26. The plant initially shutdown on 2/20/76. (LIJ)

3/17/76 "Met Ed to be Fined $8,000 by Agency."
The NRC recommends that Met Ed be fined $8,000 for a breakdown in its protection of the TMI nuclear power station against intruders. (LIJ; also3/18/76 HP)

3/29/76 "3 Mile Island Refueling Lags."
Mechanical problems will extend the refueling process to 4/30. (LIJ)

3/30/76 "U.S. Proposes to Fine Utility for not Keeping Unstable Ex-Employee out of Nuclear Plant."
The NRC proposes to fine Met Ed $8,000 for failing to apprehend a disturbed former employee who entered protected area around TMI. The fine would be the 16th such penalty for inadequate security in the last two years. (NYT)

4/14/76 "TMI Unit One Set to Start Up on May 8th."
TMI Unit I has been shutdown for seven weeks for refueling, but is set to restart on 5/8. Mechanical problems have delayed fuel replacement. (MPJ)

4/19/76 "Nader Group Says Security is too Lax."
John Abbott, of the PIRG, is asking the NRC to order beefed-up security at the TMI nuclear power plant. (LIJ)

5/4/76 "Complete Refueling."
The refueling process is completed and TMI will resume operation on 5/21. (LIJ; also 5/5/76 MPJ)

5/13/76 "Public Due Rebates, House Tells Utilities."
State House passes a bill which requires public utilities to make rebates if they are found to be overcharging on fuel adjustment. (LIJ)

5/21/76 "Nuclear Plant On Line Next Week."
An additional five or six days will be needed before the plant at TMI resumes full operation. (LIJ)

5/31/76 "Area N-Plant Tighten Guard."
Security at TMI is beefed-up in the wake of an alert ordered by the NRC. (LIJ)

6/2/76 Editorial: "Reassurance: Nuclear Alert is Encouraging."
A low level alert put out by the NRC last weekend ought to reassure the lay person that the NRC has its ear to the ground and will order extra precautions. The editorial raises a question concerning possible nuclear attack or bombings in the future and sees it as a threat. It concludes that a plan must be ready if we face such a threat. (HP)

6/7/76 "3 Mile Island back in Operation."
TMI resumed full service on 6/1/76, 13 weeks after being shutdown. (LIJ)

7/1/76 "Connors at Nuclear Drill."
Dr. Connors attended a radiation emergency drill at TMI. The drill dealt with various aspects of a hypothetical shutdown of operations resulting from

a leak of reactor coolant and an ensuing fire in a valve control center. (LIJ)

7/8/76 "3 Mile Island Unit Output Stays 100%."
For the fourth consecutive week following refueling, Unit I has operated at 100% power. One event merited reporting to the NRC; a liquid effluent monitor failed to close after a planned release of low level radioactive liquid. (MPJ)

8/18/76 "3 Mile Island Nuclear Plant Ranked First in United States."
TMI ranked first in the U.S. and eighth in the world among reactors capable of generating 150 megawatts and above. (MPJ)

"Metropolitan Edison Co."
Met Ed will have spent $169 million for environmental protection measures at generating stations including TMI by the end of 1976.

8/20/76 "Met Ed will Form Security Force."
Met Ed revealed that it will form its own security force to guard TMI.

8/25/76 "Leak at TMI Reduces Power 50 Percent."
Power at Unit I was reduced to 50% on 8/18 to correct a leak in a main condenser. A new staff analysis on environmental impact will be undertaken. (MPJ)

9/15/76 "TMI Completes Second Year of Operation."
TMI completed its second year of operation on 9/2 with an overall availability factor of 75.7%. Only one event in August, a liquid leak from a liquid waste disposal system, required reporting to the NRC. (MPJ and LIJ)

9/16/76 "Meetings Set on N-Plant Unit License."
Meetings concerning a federal operations license for Unit II of TMI will be held 9/23 and 9/24. The meetings will

consider highly technical reviews of safety considerations and be conducted by the NRC ACRS. (HP)

9/29/76 "Move to Shut Down 3 Mile Island Would Mif GPU Users."
Petitions by the ECNP to shutdown Unit I and halt construction of Unit II would raise GPU costs by $9.5 million per month according to GPU. (MPJ)

10/1/76 "3 Mile Island Reactor Plan is Rated Safe."
NRC's ONRR says the design of Unit II, planned for TMI in 1978, shows that the plant can be operated "safely without undue risk to the health and safety of the public." (HP)

10/6/76 "3 Mile Island Unit at 100% Power."
TMI Unit I is operating at 100% power. Two incidents at the end of August were reported to the NRC. (MPJ)

11/6/76 "Security Study Asked at TMI."
The ACRS of the NRC has urged that "further review" of the security measures at TMI be undertaken to reduce the possibility of sabotage. It indicated that Met Ed has been fined twice for security violations. (LIJ)

11/12/76 "Small Fire Hits TMI Pre-Fab Office."
A small fire occurred at a pre-fab construction office. Confusion reigns as firemen from a local township are denied access. (LIJ; also 11/13/76 HP)

12/28/76 "Radioactivity Found at Nuclear Plant."
Met Ed reported that a "concentration of several radioactive isotopes is 10 times higher than normal background levels." The company is confident that these above average values were due to the atmospheric nuclear weapons test recently conducted by the Chinese. (LIJ)

2/18/77 "Refueling Set for Nuke Plant."
Met Ed says refueling will begin on 3/19. The plant will be closed for 6 weeks while 61 fuel assemblies are replaced. (LIJ)

2/23/77 "NRC Plans to Create N-Plant 'Fortresses.'"
The NRC orders atomic power plants to take increased precautions to thwart armed sabotage attempts. Met Ed will bring the TMI plant into compliance with regulations. (HP)

2/24/77 "2 Area Nuclear Plants Study Tougher Security Regulations."
Two area nuclear plants (TMI and Peach Bottom) announced that they will comply with new beefed-up security regulations being issued by the NRC. (LIJ)

4/9/77 "Refueling at TMI Nuclear Plant Proceeding."
The refueling of Unit I is proceeding on schedule. The plant has been closed for 3 weeks and is expected to return to service on 5/1/77. (LIJ)

4/9/77 "In Case of Emergency at Nuclear Plant it Would Take 3-6 Hours to Evacuate 5 Mile Area."
Comments made at hearings on the license application for Unit II. The article details testimony on emergency evacuation plans and the expected amount of time to implement evacuation. (HP; also 4/12/77 LIJ)

4/20/77 "Nuclear Stations Public Hearings are Scheduled."
Refueling of Unit I continues on schedule. NRC hearings, relative to the licensing of Unit II, are scheduled for May. (MPJ; also 4/23/77 LIJ)

4/27/77 "Citizens to Protest Nuclear Radiation from 3 Mile Island."
Concerned citizens will meet at the

square in Goldsboro on 4/30 to release helium filled balloons containing cards. The balloons will follow the path of air-borne nuclear fallout from the TMI facility. Also, a discussion aimed at seeking a moratorium on nuclear power will be held at the Friends Meeting House in Harrisburg. (MPJ)

4/27/77 "3 Mile Island Flooding Causes Delay in Refueling."
Flooding in the pump house may lead to an extension of TMI Unit I refueling outage. (MPJ)

5/4/77 "Nuclear Hearings on TMI Delayed."
Hearings on Unit II were postponed because two of the three board members could not be present. (MPJ)

"Service Schedule Delayed at 3 Mile Island."
Detailed inspection of steam generators and hydraulic sway suppressors has caused a 10 day delay in the startup of TMI Unit I. (MPJ)

6/1/77 "Three Mile Island Security Up: Nuclear Station Rated as Defensible."
A detailed account of the changes at TMI relative to security against possible saboteurs. Steps were taken to conform to new NRC regulations. (HP)

6/7/77 "2 Problems Noted at TMI."
Met Ed reported two problems; both were immediately corrected and were not a "threat to the health and safety of the public." (LIJ)

6/8/77 "Health Questions Addressed Reactor Hearings Delay Requested."
Chauncey Kepford of the Harrisburg CSE asks for a delay in hearings on Met Ed's request to operate Unit II. He says time is needed to review the 5/21 testi-mony of Dr. Reginold L. Gotchy, physicist,

who said that the "nuclear fuel cycle is considerably less harmful to man than the coal fuel cycle." Statements by the NRC are detailed relative to the hearings and Met Ed's reaction to the request for a delay. (HP)

6/9/77 "Met Ed Nuclear Reactor Hearings are Extended."
Intervenors against the licensing of Unit II were given an extra day or longer to present their case. Details of the previous day's hearing are provided. (HP)

6/11/77 "Reactor Plan Doubts Voiced."
Comments of area residents at a licensing hearing on Unit II were presented. (HP)

6/16/77 "Met Ed Hearings on Reactor Have Been Extended."
Hearings on Unit II licensing have been extended in order to provide citizens with additional time to voice objections. (MPJ)

6/16/77 "Wastes from Three Mile Island: Danger From Radiation Leak Denied."
A description of the incident involving leakage of radioactive liquid from a truck hauling waste from TMI is presented. State BRP spokesman compares the level of radiation to that of a radium dial watch. (HP)

7/13/77 "Public Hearings were Completed."
Hearings on the Unit II operating license were completed on 7/5 in Bethesda, Md. (MPJ)

8/10/77 "Tighter Security Measures Rejected at Nuclear Sites."
The NRC has rejected a PUC petition seeking tighter security measures at nuclear plant construction sites adjacent to operating units. Security breaches in 1975 and 1976 at TMI were part of the argument for tighter security. (MPJ)

8/14/77 "Nuclear Power Lags While Foes Flourish." Richard Pollock (Critical Mass) says construction slowdown in the nuclear power industry may be due more to financing problems, safety problems, construction delays, and operating efficiency problems than to organized opposition. (NYT)

9/19/77 "3 Mile Island Nuclear Station Out of Service." The TMI nuclear generating station will be out of service until Tuesday due to a chemistry problem in the steam generator feedwater system. (LIJ)

11/23/77 "Female Engineer at TMI in Challenging Career at Nuclear Plant." A feature story on Linda A. Fisher, the only female engineer at TMI. (MPJ)

"TMI Alert Group Holds Public Meeting." A report on a meeting of the TMIA. (MPJ)

11/26/77 "Nuclear Unit Lists Radioactive Leak." Met Ed reported to the NRC a leak of radioactive gases. The leak was immediately corrected. It was not considered a threat to the health and safety of the public. (LIJ)

12/7/77 "TMI Alert States Second Public Meeting." A second public meeting of the TMIA group will be held. (MPJ)

12/14/77 "Three Mile Island Report." Unit I is operating at 100% power. No events were reported to the NRC this week. Unit II construction continues. (MPJ; also 12/20/77 LIJ)

1/11/78 "Radioactivity Found Near TMI." Met Ed reported to the NRC of increased radiation in the sediment in the Susquehanna River one mile away from the plant. Met Ed claims a radiation increase is due to Chinese testing of nuclear explosions. Met Ed also claims that the radioactive gases' concentrations were very low and "resulted in negligible exposure to the public." (LIJ)

1/17/78 "Emergency Evacuation Plan Discussed."
The Dauphin County CD Director reviews a plan for evacuating the area surrounding the TMI nuclear power station. The meeting was sponsored by TMIA. (HP)

2/1/78 "TMIA to Meet."
TMIA will sponsor a public meeting on the monitoring of TMI by Pa.'s DER. (MPJ)

2/13/78 "Met Ed to Load Nuclear Fuel at TMI."
Met Ed has received an operating license for Unit II at TMI from the NRC. Unit II will immediately be fuel loaded for operation. Unit I provides about one-third of all electricity generated by Met Ed. (LIJ)

2/15/78 "Monitoring Radioactivity in the Environment."
The monitoring of radioactivity in the environment was the subject of a TMIA meeting. DER representatives discussed how this is done. (MPJ)

"TMI Unit 2 Receives License."
NRC issued an operating license for Unit II on 2/8. A brief history of TMI and licensing procedure was summarized. (MPJ)

3/2/78 "Start-Up of Three Mile's Unit 2 Stayed."
The U.S. Circuit Court of Appeals orders a temporary halt to the activation of Unit II. Chauncey Kepford claims that the NRC failed to notify citizens groups that a license had been granted. (HP)

3/4/78 Editorial: "Nuclear Power: Time to Resolve a National Issue."
The editorial alleges that the NRC is in the business of assisting the development

of nuclear power plants. Hearings relative to licensing do not solve the real issue, i.e., whether our country should be pursuing a policy of nuclear development. It raises a number of questions such as what is a safe level of radiation, how should radioactive waste be disposed, what will be done to plants after useful life is expended, and is nuclear power the only alternative to our energy crisis. The editorial then calls for an independent congressional inquiry and appraisal of the risks and benefits of nuclear power. (HP)

3/8/78 "TMIA Meeting."
TMIA will meet to review its previous two meetings. (MPJ)

"Second Unit Approved for TMI."
A circuit court gives approval for a sustained nuclear reaction in TMI Unit II. This follows appeals of environmental groups. (MPJ; also 3/9/78 HP)

3/24/78 "Stay of NRC's Nod for Unit 2 Requested."
Chauncey Kepford asked the NRC Appeal Board to shutdown Unit II and reopen public hearings on Met Ed's request for operating authority. Kepford says Randon 222 is underestimated as a potential health hazard. (HP)

3/30/78 "TMI Second Unit Reacts."
Met Ed announced that the Unit II experienced a chain reaction in the reactor core. It is expected to go on line in late May or June. (LIJ; also 4/5/78 MPJ)

4/1/78 "Bid to Stop N-Reactor Rejected."
The NRC Appeals Board sees no reason to suspend Unit II operations. Harrisburg CSE and York CSE filed objections charging a major source of radioactive emissions in the plant's uranium fuel cycle is underestimated by the government. (HP)

4/5/78 "The Press and Journal Asks."
A roving reporter asks eight people if they feel threatened by the closeness of TMI. Most reply "no." (MPJ)

4/26/78 "TMI Environmental Report Filed with Borough Library."
A 61 page report listing all radiological data in a 20 mile radius of TMI was delivered by Met Ed to the Middletown Public Library. (MPJ)

5/2/78 "Full Reactor Use May Be Delayed."
Unit II will probably not be brought to 100% power this month. The plant became radioactive on 3/28. It was shutdown for the fourth time on 4/23 when a safety device was set off by instrument noise. (HP)

5/10/78 "TMI Reports Made."
The article reports on the refueling of TMI Unit I and maintenance of Unit II. Also, two events which occurred during testing are reported to the NRC. (MPJ)

5/31/78 "TMI's Unit 2 Lags on Production."
Unit II will not produce electricity at full capacity until late July or early August according to Met Ed. (HP)

6/1/78 "TMI Reports."
The NRC gave TMI permission to operate Unit I at 100% of full power. Extensive testing of Unit II turbines is underway due to recent problems. Two events were reported to the NRC in the previous week. Neither event is said to have any possible effect on the health and safety of the public. (MPJ)

6/7/78 "TMI Reports."
Unit I is at 100% power. Unit II is being readied for cold shutdown in preparation for inspection of a vessel head. One event required reporting to the NRC in the previous week; one of

two emergency diesel generators did not operate for the full length of the test. The incident was termed as not harmful to the safety or health of the public.

6/14/78 "TMI Nuclear Station Report."
Feed pump auxiliary condenser leaks cause a reduction of the capacity of Unit I to 96%. Inspection of Unit II's vessel head continues. A discrepancy in the support provided by a relief valve required reporting to the NRC. (MPJ)

6/21/78 "TMI-I Under Full Power."
Unit I is at full power and Unit II inspection continues. One event was reported to the NRC. A pressure switch was set at greater value than was required by specifications. (MPJ)

7/6/78 "TMI Reports."
Unit I is expected to be back on line following work on reactor coolent pump. Engineering design work for modifying steam valves in Unit II continues. One event was reported to the NRC. A short circuit caused a control fuse to blow. (MPJ)

7/15/78 "Stranded Boater Climbs Fence Into N-Plant."
A stranded boater caught security personnel at TMI off guard when he climbed a fence topped with barbed wire, squeezed through a gate, knocked on two doors, and finally surprised two guards. A Met Ed spokesman said the "incident was not considered a breach of security because the boater was not in a sensitive area." (LIJ)

7/18/78 Editorial: "The Met Ed Fiasco."
The editorial urges a full scale investigation into Met Ed's fitness to operate a nuclear power plant due to the "stranded boater incident." (LIJ)

7/21/78 "Nuclear Station on Full Power."
Unit I is operating at full power. Work on piping modifications for installation of new main steam relief valves for Unit II has begun. (MPJ)

7/25/78 "License Won't be Suspended. TMI Reactor will Continue Operations."
The ASLB finds insufficient cause to reopen the record on emergency evacuation plan. Sufficient ambiguities and inconsistencies relative to aircraft crash possibilities are not found. (HP)

7/26/78 "Citizen's TMI Tour Provides Insight of Nuclear Power."
Local government officials and civic groups tour TMI. The purpose of the tour was to increase public awareness. (MPJ)

"Security Gate Found Unlocked at TMI Station."
There was a disabled boat incident involving security at TMI on 7/13. A gate was found unlocked. The gate was left open for ten minutes; no unauthorized persons were detected. (MPJ and LIJ)

8/10/78 "Critics of Unit 2 File Risk Appeal."
Chauncey Kepford filed a brief with the NRC asking for a review of an ASLAB decision not to suspend operation at Unit II. Kepford's criticisms include the possibility of an airplane crash and inadequate evacuation plans. (HP)

8/16/78 "TMI Reports Made."
The Unit I reactor was operating at 100% full power. Unit II was being fitted with main steam safety valves. An incident reported to the NRC involved the actuation of one of the unit's fire protection systems. It was uneventful and of no consequence to the public's health and safety. (MPJ)

8/23/78 "Met Ed TMI Report."
Unit I is operating at full power. Unit II is undergoing piping modifications to accommodate the installation of the steam relief valves. An event occurring in Unit I resulted in a report made to the NRC. One emergency diesel generator was removed from service without completing the prerequisite of testing the second diesel generator. The test was subsequently performed with no complications, and the public's health and safety were unaffected. (MPJ)

8/30/78 "TMI Report is Announced."
TMI's Unit I is operating at full power. At this point, installation and testing of the new steam relief valve is near completion. An event which occurred in Unit I was reported to the NRC. Met Ed was informed that weld filler wire may have been incorporated in Unit I's construction. This material is normally not used in the construction of reactor vessels. It was determined that the possible presence of the weld filler wire did not affect the integrity of the reactor vessel. A report was also made to the NRC concerning Unit II. A missing fire barrier seal was the topic of the report. Corrections were made, and neither of the two events reported to the NRC caused public health or safety problems. (MPJ)

9/7/78 "TMI N-Station."
TMI Unit I was reported to be operating at full power. Unit II's new steam safety valves were installed, and testing was in progress. An event occurring at Unit II was reported to the NRC. A one-quarter inch hole drilled into a reactor building access corridor for bracket mounting was leaking at a rate exceeding its limit. The hole was sealed, and the leaking was of no consequence to the public. (MPJ)

9/20/78 "1.2 Billion N-Plant Dedicated in Dauphin County."
Unit II at TMI is dedicated. The principal speaker is James O'Leary, of DOE. O'Leary states that the future for nuclear power is not bright unless safety problems are corrected. (LIJ and HP)

9/27/78 "Mayor Joins Area Communities in Formulation of Emergency Plan."
The article expresses Middletown Mayor Robert Reid's concern over organizing an emergency plan in the event of a local emergency. (MPJ)

10/4/78 "TMI-2 Power Ascension Proceeds."
Unit II is operating at 40% power so that testing can be completed. Further testing at 75% power is expected. Unit I is reported as operating at 100% power. Two events at TMI were reported to the NRC. The first incident involved an excess amount of sodium hydroxide in a storage tank in Unit I. The imbalance was due to faulty equipment, and the problem was corrected. Unit II experienced problems with the emergency core cooling system. A faulty electrical component was the cause. The problem was corrected, and neither event affected public health or safety.

10/11/78 "TMI Unit 2 Continues Power Testing."
Testing of Unit II at 40% power is complete, and following repairs, more testing at 75% power will be done. Unit I is reported to have been operating at 100% full power. Three events were reported to the NRC; two occurring in Unit II and another in Unit I. None of the incidents affected the public health or safety. (MPJ)

10/18/78 "TMI Reports Listed."
Unit II is still being tested, but Unit I is operating at 100% full power. Met Ed reported three events to the NRC.

All three occurred in Unit II. Causes of the problems included a faulty damper, a malfunctioning isolation valve, and improperly enclosed electrical connections. None of the events posed health or safety threats to the public. (MPJ)

10/18/78 "Security Drill at TMI."
A security drill will be held by TMI on 11/8. The drill will involve only TMI personnel. Concern was expressed that the area emergency units are only being invited as spectators. The article points out that better coordination is needed between TMI and the surrounding community. (MPJ)

11/1/78 "Testing Continues at TMI-2."
Unit II is still undergoing testing and Unit I is operating at 100% full power. The NRC was notified of one event which occurred at TMI Unit I. An electric heating element developed a small leak. (MPJ)

11/2/78 "Jersey Agency Resists TMI Overrun."
N.J. Department of the Public Advocate said that the costs of constructing Unit II should not be passed on to N.J. and Pa. consumers. It asks that $12 million be deleted from the JCPL rate base. This is their share of alleged construction overrun costs. (HP)

11/8/78 "TMI-2 Power Level Increasing."
Unit II testing is completed, and the unit is restored to 100% power during the week of 11/6. Unit I is operating at 100% power, and no events are reported to the NRC. (MPJ)

"TMIA Sets Nov. 14 Meeting."
The article announces the date, time, and place of a public meeting of TMIA. Two speakers, Dr. Chauncey Kepford and Dr. William Lochstet are to discuss the chemical and gas emissions of TMI and

are also expected to explain the operation of a nuclear power plant. (MPJ)

11/29/78 "TMI Unit I Receives Good Rating From Inspectors."
The article reports that the TMI Unit I reactor received a good rating in terms of regulatory performance. The criteria for the rating are explained along with other NRC evaluation methods. (MPJ; also 11/30/78 LIJ)

12/12/78 "Met Ed, US Differ on TMI Over Plane Crash Chance."
A public hearing was held by the NRC to determine the chances of a jet plane crash into TMI's cooling towers. The hearing was held by the NRC's ASLAB. The hearing was called in response to a challenge to Unit II's operating license by CSE. The citizens are concerned due to jet travel from a runway at Harrisburg International Airport which is located 2.7 miles from TMI. The board refused to close TMI while this issue is heard and decided. (LIJ)

12/13/78 "Reactor 2 Resumes Testing."
TMI Unit II was being tested at 90% power, and further testing including a planned shutdown and restart at full power was expected. Unit I was operating at 100% full power. One event was reported to the NRC. It involved faulty switches in the Borated Water Storage Tank. The problem was corrected, and the public was unaffected. (MPJ)

12/26/78 "Pump Repair Routine Cleaning: Closes N-Station Unit."
Unit II is closed for feed water pump repair and cleaning of the boiler feed water purification system. (LIJ)

12/28/78 "TMI Station Reports."
TMI Unit II was shutdown for repair. Unit I was operating at full power.

Two events were reported to the NRC. Neither caused any serious complications, and the public remained unaffected. (MPJ)

1/12/79 "TMI Facility Observation Center Will Reopen."
New displays, hours, and location of the TMI Observation Center are listed.

1/17/79 "TMI-1 and TMI-2 at Power."
Unit I is at 100% power. Unit II is at 90% power. Unit II's reduction is due to a tube leak in the main condenser. (MPJ and LIJ)

1/24/79 "TMI-2 Notes Repairs Made."
Unit II is shutdown for various types of valve repairs. Failure of a pump that takes air samples to the reactor building monitoring devices was reported to the NRC. (MPJ)

1/27/79 "Scientists Advise Unit I Closing."
The UCS called for a shutdown of the Unit I reactor. The scientists say the safety system's electrical cables at TMI will fail during a fire, and the plant's safety system cannot withstand accidents it was designed to control. (HP)

1/30/79 "Radioactive Wastes Pile Up at Two N-Plants."
Both plants have to store radioactive wastes as the federal government's plan for reprocessing of spent fuel is now on hold. (LIJ)

1/31/79 "TMI Unit 2 Reports."
Unit II is shutdown with the reactor coolant system depressurized while level measuring instrument isolation valves on the pressurizer are being repaired. (MPJ)

2/14/79 "Met Ed Announces that Both TMI Units on Line."
Unit I is at 100% power. Plans are being made for refueling. Unit II is limited to 92% power until a heater drain pump and heater drain valve are repaired. Six events were reported to the NRC; one involved Unit I and five involved Unit II. None of the events were termed a threat to public health or safety. (MPJ)

"TMIA Announce Rescheduled Public Meeting."
TMIA will meet in Harrisburg. The focus will be on rate impact. (MPJ)

2/21/79 "TMI-1 Prepares for Refueling."
Unit I is at 100% power but will shutdown for refueling. Unit II is at 90% power. Two events were reported to the NRC. They concerned a tripped breaker at Unit I and the need for recalibration of switches at Unit II. (MPJ)

"Refueling of Unit 1 at TMI Proceeding Along Guidelines."
Refueling of Unit I is to last from 2/17-4/2. The refueling operation is described. (MPJ)

"Billion Dollar N-Plant."
A 10 year billion dollar effort was capped with the opening of Met Ed's nuclear power plant at TMI in Sept. (MPJ)

2/22/79 "Cites 'Costly Electricity.' Foe Zap 'N-Waste.'"
George Boomsma, antinuclear power leader, speaks at a TMIA meeting. He refers to the uneconomics of nuclear power and predicts a rise in electric utility bills. (HP)

2/27/79 "Harold Denton of NRC Testifies Before Cong. Udall's Congressional Comm."
The risk of a serious reactor accident

is "acceptably low." Cong. Udall is Chairman of the House Energy Committee. (NYT)

3/1/79 "Refueling of N-Plant."
Unit I is being refueled for the 3rd time since 1974. The plant is closed during refueling. (LIJ;also 3/7/79 MPJ)

3/14/79 "Manager of TMI is Announced by Met Ed."
Gary P. Miller was named manager of the TMI nuclear generating station. (MPJ)

3/21/79 "TMI-2 at Full Power."
TMI Unit II is operating at full power, while Unit I is still out of service due to refueling. An event involving a faulty high pressure injection pump was reported to the NRC. The problem was corrected, and the public remained unaffected. (MPJ)

3/22/79 "N-Plant Plane Hazard Hearings Set."
Hearings are scheduled to determine whether aircraft larger than 200,000 pounds should be allowed to take off or land over TMI. (HP; also 3/28/79 MPJ)

3/28/79 "Unit One Refueling Completion is Near."
Preparations to restart Unit I after refueling continue. Unit II is operating at full power. Two events were reported to the NRC. Both involved the accidental tripping of diesel generators. The disorders responsible for the two events were corrected, and the public's health and safety were unaffected. (MPJ)

3/29/79 "Accident at TMI N-Plant Near Harrisburg, PA; Releases Above Normal Levels of Radiation into the Atmosphere on 3/28."
The exact cause of the TMI radiation release is unknown. The radiation release is not a threat to the people of the area. The release of radiation into the containment building is not viewed as endangering the health of workers at TMI. (NYT)

3/29/79 "Disagreement Between Met Ed and B&W Over Valve Failure as Cause of Accident." James Higgins, NRC staffperson, said the cause of the accident is a series of filters that did not operate properly. (NYT)

"N-Mishap at TMI Spills Radiation Over 16 Miles."
Gary Snyder, NRC official, stated that the accident was "probably the biggest radiation leak," but he quickly added that the safety record of the nuclear industry was excellent and there was no threat to the public. (LIJ)

"Nonchalant Reaction Over Leak."
People within the shadow of TMI are not concerned over it. (LIJ)

"Middletown People Starting to Worry."
Interviews with some residents show concern over accident and perceived threat to them. (LIJ)

"Three Things that Shouldn't Happen Did."
A key valve that should have stayed open closed and two leaks developed which should not have occurred. (LIJ)

Editorial: "How Close Were We to a N-Catastrophe."
The editorial demands that Met Ed explain why the accident occurred and prove that the plant can be operated safely. Until then, TMI should remain closed. (LIJ)

"Timing of Mishap Bad for N-Industry."
Timing of the accident occurred when the industry is under severe criticism from such groups as the UCS and Critical Mass. (LIJ)

"Goldsboro Takes Crisis in Strides."
Residents of Goldsboro are not concerned about TMI's impact. (LIJ)

3/29/79 Editorial: "N-Accident. No Time for Secrecy."
This editorial makes a plea for telling the truth and telling it promptly when something goes wrong at nuclear power plants. (HP)

"In Initial Reports of Radiation Leakage; Scranton: Emission Levels were Played Down."
Lt. Gov. Wm. W. Scranton III contends Met Ed underplayed the amount of radioactive emissions. The article comments on possible additional emissions and reports details of the accident as well as delays by Met Ed. It also gives DER reaction to the situation. (HP)

"N-Plant Proponents See Fuel for Critics."
Reactions of pro and con nuclear groups are updated. Previous problems at TMI are repeated. It is expected that antinuclear groups will use the crisis as a rallying point. (HP)

"Governor's Home Scene of Protest."
Antinuclear group holds a protest outside the governor's home. Thornburgh says TMI is under control. The group believes otherwise. Comments by TMIA. (HP)

"He Favors N-Power Despite Accidents."
Energy Secretary James Schlesinger comments on the good safety record of nuclear power, its contribution to the economy, and role in the fuel crisis. (HP)

"N-Leak Explanation Demanded."
Congress demands an explanation of the previous day's accident at TMI. Cong. Ertel, from the district where TMI is located, states he recently questioned the safety of TMI. Reactions of various congressmen are provided. (HP)

3/29/79 "Hospitals Can Test Those Linked to I-131."
Local hospitals can test people exposed to the Iodine 131 leak at TMI. A hospital spokesman discusses the element and relation to thyroid problems. (HP)

"Radiation Being Vented; Delay in Alert Assailed. 3 Mile Sealed Off."
A report on the accident of 3/28 and continued venting of radioactive material. It gives initial reactions of Met Ed and the NRC and lists various groups brought in to respond to the situation. Also discussed was the delay in reporting the accident to local officials. (HP)

"Melt-Down Most Feared of Accidents."
The article explains how cooling systems work in nuclear reactors and what happens if they breakdown. (HP)

"Protest Filed Over Delaying Announcement."
York County CD director lodges a protest with PEMA. He claims that TMI delayed from 4 AM to 7:27 AM to notify him of the breakdown. (HP)

"Goldsboro: Tranquility and Anger."
Reactions of Goldsboro's officials and residents about the accident and delay of reporting it. (HP)

"Royalton Never Got the Word."
The article indicates that Royalton Borough, three miles from TMI, was not officially notified of the leak. It also tells of the delay in getting information to Middletown Borough. (HP)

"Call for Investigation. Area Officials Concerned Over Proper Notification."
Cumberland County Commissioners call for an investigation as to whether "timely notice" was given to county officials relative to the previous day's accident. (HP)

3/29/79 "Radiation's Two Sides: Boon, Bane." This is a discussion of types of radiation and its effects on the human body. (HP)

"N-Plant Accidents Chronicled." A.P. story on 10 nuclear accidents in the U.S. and abroad from 9/76 to 3/28/79 at TMI. (HP)

"N-Plant Gone Haywire: Good Idea for Movie, but Real-Life Quandary." Discussion of the movie *China Syndrome*. It asks whether drama could happen and documents how parts of it have occurred in reality. (HP)

"Unit Two Center of Controversy. TMI: Legacy of Trouble." Details protests and troubles at TMI in regards to Unit II from its construction on the island in 1967 until the present time. (HP)

"Winds Keep County's Dose to a Minimum." Winds are keeping Lancaster County's exposure to the radiation at a minimum. (LIJ)

3/30/79 "NRC Chairman Joseph Hendrie Testifies Before Congressional Comm." Failure of several safety systems caused the accident. The record shows similar equipment failure at TMI during 1978. (NYT)

"Radioactivity Continues to Leak From Crippled TMI Plant." The Gov. says there is no cause for alarm. Met Ed discloses at a Congressional hearing that 180 to 360 of 36,000 fuel rods melted. Antinuclear groups mass protest. (NYT)

"Mayor Reid of Middletown Stated That Met Ed Delayed Notifying Town of Accident." The Company's evacuation plan is poor and Met Ed officials play down danger. Antinuclear scientists like Drs. George Wald and Ernest Sternglass warn people against danger. (NYT, LIJ, and HP)

3/30/79 "Radiation-Who Said What." The article chronicles announcements by various government, Met Ed, and other officials concerning radiation as a result of TMI accident. (HP)

"Lawmakers Privately Briefed at TMI." The article gives details of a visit by four U.S. senators and five congressmen to TMI and their reactions to the accident. (HP)

"Evac. Plans Due More Work." Cumberland County evacuation plans are discussed. Plans are under development, but not complete. (HP)

"Schlesinger is Cautioned." Sen. Ted Kennedy (Mass.) asks James Schlesinger to reconsider a bill speeding nuclear power plant licensing. He also proposes safety measures. (HP)

"Students Measure Extent of Fallout." Students take a field trip to measure TMI radiation fallout and see how detectors work. (HP)

"Foul-up Sketched for Panel." Federal regulators present information on TMI accident to concerned members of Congress. (HP)

"N-Spill 'Could Happen Again,' Met Ed Says 4 Counties Still on Alert." Dauphin, Cumberland, York, and Lancaster Counties remain on standby evacuation alert. Met Ed spokesman says another accident "could conceivably happen again." Details provided on what happened inside the reactor during the accident. It also contains other comments by local government and Met Ed officials. (HP)

3/30/79 "N-Dangers Doubted."
The NRC contends there is 'no danger' to people residing in the off-site area. Gov. Thornburgh encourages people not to disrupt daily routine. A report on decreasing radiation levels in Goldsboro is included. (HP)

"Poll Shows Middletown Uneasiness."
A report on a survey conducted by Media Statistics Inc. from Silver Springs, Md. The poll centers on whether people would leave the area if a repeat accident occurs and their perception of danger. (HP)

"Low-Level Radiation Still Leaking from TMI Plant."
Radiation continues to leak from Unit II of TMI in what experts say is probably the worst accident in the history of commercial nuclear power. Met Ed and government officials state that the radiation posed no hazard to the public and no nearby residents have been evacuated. (LIJ)

"Mishap Gets Major Coverage Overseas."
The story of foreign newspapers' reaction to the TMI accident.

"Walker Tours Plant; Nuke Faith Unshaken."
Cong. Walker said that the nuclear accident at TMI has not shaken his faith in the safety of nuclear power, although he feels it may indicate that some safety design improvements are needed. (LIJ)

3/31/79 Gov. Thornburgh Advises Pregnant Women and Children Within 5 Miles of TMI to Leave."
Met Ed's venting releases a high amount of radioactive gases. The NRC states no immediate danger to the public. Maybe 9,000 rods melted. (NYT, LIJ, and HP)

3/31/79 "Streets of Goldsboro Empty."
A worse fright comes to Goldsboro when an order to evacuate schools within 5 miles of TMI was received. (NYT and HP)

"NRC States Remote Risk of Meltdown Exists."
The NRC claims there is a remote risk of a meltdown due to a large pressurized bubble of hydrogen gas that has formed in the top of the reactor's vessel. GAO reports that evacuation plans for areas surrounding most nuclear power plants are inadequate to protect the public in a serious accident. (NYT, LIJ, and HP)

"Energy Secy Schlesinger says Nuclear Power Continues to be Essential."
Despite the TMI accident, Schlesinger says nuclear power is essential. (NYT)

"No Dangerous Radiation Level Here, He Says, City's Mayor Issues Appeal for Calm."
Harrisburg's Mayor, Paul E. Doutrich, Jr., meets with Met Ed officials and says there is no dangerous radiation in the city. Mayor details plans and calls for calm. (HP)

"Radiation, Who Said What."
Conflicting reports from utility, federal, and state authorities on radiation. (HP)

"Conflicting Statements, Sifting Facts, Fiction Proves Formidable Job."
This article examines the reports of fluctuating levels of radiation and possible dangers. (HP)

"Carter's Emphasis on Safety."
Pres. Carter comments on TMI and implications for stricter safety measures in the future. (HP)

3/31/79 Editorial: "N-Danger. An Intolerable Price to Pay."
The editorial contends that the fear and disruption of normal life that resulted from the TMI accident is unacceptable in a stable, functional society. It discusses gaps in knowledge, research, costs in nuclear power, and money man-power expended. It concludes that the cost is too high a price to pay. (HP)

"All Quiet in Bainbridge, Falmouth."
People in Bainbridge and Falmouth are taking the accident seriously. Traffic is minimal. Most homes are closed up, windows shut, and curtains drawn, despite temperatures in the high 70's. (LIJ)

"U.S. Experts take Charge of Crisis at Plant, 'No Immediate Danger.'"
Harold Denton, of the NRC, is in charge at TMI. The accident is the "most serious in our licensed program," but there is "no immediate danger." (LIJ)

"N-Plant has a History of Problems."
NRC records show that TMI had to be shutdown on four other occasions after mechanical problems in the last 12 months. (LIJ)

"People Near TMI Pack, Leave."
Schools are closed within the five mile area of TMI; people are packing and leaving. (LIJ)

Editorial: "A Battle Against the Unknown."
The editorial argues that the residents of the immediate area near TMI should not support a restart of TMI until assurances can be given of no future harm. (LIJ)

3/31/79 "York Students Bused by Mistake. A Long Day to be Remembered."
Students from Red Lion School and Fishing Creek Elementary School were evacuated by mistake. (HP)

"West Shore Pupils Evacuated."
West Shore Schools close; pupils are taken to Northern York. Parents of other students remove them from school. (HP)

"Met Ed Executive Says: Reactor will be Dismantled."
Discusses temperature inside the reactor, damage to fuel rods, and belief that Unit II will have to be dismantled. Met Ed details sequence of events in the accident. (HP)

"Congressmen Deluged by Calls. 2 Senators Act to Aid in Alert."
U.S. Senators Schweiker and Heinz react to the NRC and TMI. White House Task Force is set up to cope with the crisis. (HP)

"Met Ed VP Disputes State on its N-Leak Data."
John Herbein takes exception to information on TMI provided to the public by the State. The article focuses on venting and radiation. (HP)

"Many Schools on East Shore Dismiss Early."
Details school closings and plans to use buses if the emergency worsens. (HP)

"People Were Calm. Police Force Beefed Up in Middletown Curfew."
A curfew and extra police protection in Middletown is provided. The article presents residents' reactions. (HP)

"Staying Facility Ready."
The State Farm Show Building is being readied as a staging area or evacuation

center. Hersheypark Arena is also a possible site. (HP)

3/31/79 "A Sort of Non-British Stiff Lower Lip. Calm Harrisburg Strides Through Alert." The article details how Harrisburg maintained normal routines during the crisis. Anecdotes on local people are provided. (HP)

"An Unfolding Drama. Arena Becomes Home for 'Nuclear Refugees.'"
Hersheypark Arena becomes home for 154 evacuees from the TMI area. The article gives details of the evacuation order for pregnant women and children. It tells of events at the arena and medical availability. (HP and LIJ)

"Cumberland Asks Expansion of 5-Mile Radius for N-Accident Evacuation Planning."
Cumberland County recommends expansion of nuclear accident evacuation area from 5 to 10 miles. (HP)

"County Preparedness Director has Big Job 'Trying to Keep the Whole County Calm.'"
A report on the activity of the Dauphin County Director of Emergency Preparedness, county links to information sources, and notification gaps. (HP)

"Reactor Cooling Course Pivotal."
An NRC spokesman is quoted on the need to decide how to cool down Unit II. He blames operators for the release. Agreement of Met Ed and the NRC on the situation is noted. (HP)

"GAO: States Feeble on Radiation Crisis."
GAO concludes that 43 states where nuclear reactors are located are ill-prepared to cope with radiation crisis. (HP)

4/1/79 "Accident at TMI Causes Nuclear Stocks to Decline on NYSE."
Stocks of companies in the nuclear power industry declined on the stock market. (NYT)

"H. Denton says Measures to Cool Reactor at TMI are Working."
Denton expresses doubt that Met Ed's assertion that the hydrogen bubble has been significantly reduced is valid. Joe Hendrie, of the NRC, states a 20 mile precautionary evacuation may be necessary. Pres. Carter will visit TMI. (NYT)

4/2/79 "Accident at TMI Jeopardizes Nuclear Solution to US Energy Problem."
The viability of nuclear power as part of U.S. energy independence is now questioned. (NYT)

"Pres. Carter Visits TMI."
Pres. Carter urges people to cooperate with Gov. Thornburgh. Denton reports the size of the bubble is decreasing at press conference with Pres. Carter present. (NYT, LIJ, and HP)

"Travel Aid Numbers are Listed."
The article gives telephone numbers for persons with no transportation or special transportation needs in the event of evacuation from certain Dauphin County towns. It also makes reference to upstate counties' (Union, Snyder, and Northumberland) development of plans to receive potential evacuees. (HP)

"Evac. Plans."
This article is subdivided with write-ups for Dauphin, Cumberland, York, and Perry Counties. Also included is an announcement of contingency evacuation plans. (HP)

4/2/79 "Radiation Hazards Explained."
The story focuses on how isotopes affect health. It also discusses the process of decay of radioactive elements. (HP)

"Area Cows Kept Indoors as a Precaution for Milk."
Area dairy farmers keep cows indoors. The cows are eating stored food to prevent exposure to radiation contamination. Pa. Agricultural Secretary refers to zero level radioactivity in the area. (HP)

"On Reactor Problems Plant Report 'Ignored.'"
The NRC rejected recommendations last January concerning potential problems in maintaining pressurizer-level indicators during transients in some B&W plants including TMI. (HP)

"Some Answers to Questions on N-Material."
Questions and answers from National Geographic on various aspects of nuclear energy and uranium. (HP)

"N-Energy Opponents Rally."
The article contains references to reactions to nuclear energy by protestors and antinuclear organizations across the country in the wake of TMI. (HP)

"2 Congressmen Urge Closing of Eight Similar Reactors."
Sen. Hart (Colo.) and Rep. Udall recommend closing or reducing the output of all reactors made by B&W. (HP)

"Gov. Brown asks Closing of N-Plant Near Sacramento."
California Governor asks the NRC to close Rancho Seco plant near Sacramento as a precaution. The plant is a near duplicate of TMI. (HP)

4/2/79 "Defensive Builder Insists: TMI Plant 'Safe.'"
The article provides information on B&W and their operations in Lynchburg, Va. (HP)

"City Keeps Tabs on Homes Left Behind."
Police compile a list of houses in Harrisburg from which owners have evacuated. The Mayor urges calm and sees no current threat to the city. (HP)

"Trains, Buses Take Exocus."
Information of heavier than usual use of buses and trains leaving Harrisburg. (HP)

"Evacuate: It Would be a First."
American Red Cross comments on evacuation. Earlier large scale evacuations were duc to hurricanes. Some details on such evacuations are presented. (HP)

"Waste Disposal on a Giant Scale Planned."
Photo story. Flatbed trucks are to be used to carry dangerous nuclear waste from TMI for disposal. (HP)

"Those Going Aren't Stopping Short."
Accounts of where some Harrisburg people are going to evacuate.

"N-Bunker is Under Capitol."
If the TMI area is evacuated, Gov. Thornburgh will retreat to a radiation proof bunker under the State Capitol. It provides a description of the facility. (HP)

"Lead Bricks for Radiation Shield Arrive."
The photo story shows a cargo of lead bricks to help contain radiation at TMI. (HP)

4/2/79 "Most Schools Closed."
Most schools in the Harrisburg area remain closed despite Thornburgh's recommendation that only those within 5 miles of TMI shutdown. (HP)

"Bubble Answer Needed in 5 Days."
A decision is necessary relative to removing the gas bubble from Unit II. Possible procedures for removal, people's fears about explosion, and continued radioactive releases are discussed. (HP)

"General Absolution is Granted."
Several Catholic Churches give General Absolution to members. The article discusses what this means relative to forgiveness of sins. Church attendance is down at both Catholic and Protestant Churches. (HP)

"Exposed Met Ed Worker Back on Job."
A Met Ed worker exposed to 4 rems of radiation while recovering a sample of reactor coolant is back to work. His feelings, and accounts of other exposures are provided. It discusses how radiation exposure is measured. (HP)

"Moving Newborn Eyed."
Contingency plans to move 46 newborn babies from Harrisburg area hospitals. (HP)

"Lawmakers will Gather Energy Data."
Congressional attention turns to nuclear related matters as a result of TMI. There is a call for hearings. (HP)

"N-Power Future in Doubt."
How TMI might become an impact on the future use of nuclear energy. (HP)

4/2/79 "Mayor Wohlsen Issues Release of City's Plans."
The Lancaster Mayor released a statement indicating various activities of the city. He also indicated that "contingency plans for evacuation of an area of the county extending outward from the plant for a radius of 20 miles are being put together." (LIJ)

"110,000 in County Could be Evacuated."
Precise preparations for the possible evacuation of as many as 110,000 people in Lancaster County in a 12-hour period were made Sunday afternoon in meetings at the county courthouse. (LIJ)

"From Pulpit, Pastors Face Unknown."
Facing the unknown was a common theme of area pastors in Sunday sermons. (LIJ)

"Plans for Evacuation Readied-Just in Case."
Lancaster County officials continued to draw up evacuation plans even as the immediate hazards from TMI diminished. PEMA is coordinating evacuation plans. (LIJ)

"Protesters, Pa. Agencies Mobilize."
PEMA and the National Guard become active in wake of the TMI crisis. Individual protestors are at the capital building. (LIJ)

"Farmers Face Big Decisions with Animals."
The question of care of animals in the event of TMI evacuation is addressed. Pa. DOA is involved in sampling milk, water, and feed supplies near TMI for health risks. (LIJ)

"Physicist Downgrades Accident Possibility; Udall Disagrees."
Disagreement exists between Cong. Udall and Dr. N. Rausmussen, MIT Physicist Professor, over the seriousness of the

accident. The former says a threat still exists, while the latter discounts a threat possibility. (LIJ and HP)

4/2/79 "2 Hospitals Near N-Plant Short of Personnel."
Harrisburg Hospital and HMC report a shortage of hospital staff. (LIJ)

"River Water Drain for Drinking A OK."
Testing by Lancaster City indicates that radioactive gases still leaking from TMI apparently have not contaminated drinking water drawn downstream. (LIJ)

Editorial: "America's Full Resources at Work on our Problems."
The progress at TMI in controlling the hydrogen bubble gives hope that a nuclear catastrophe can be avoided. (LIJ)

"The Dream Turns into a Nightmare."
The article describes how the safety systems failed to operate causing an accident which officials said could not occur." (LIJ)

"600,000 May be Evacuated if Necessary."
Plans are described for a 20 mile radius evacuation if necessary. (LIJ)

"Police Disperse N-Protest."
TMIA group is dispersed from the capitol area, but vow that they will return. (HP)

"Odd Fellows Move Patients to Harrisburg."
Odd Fellows Home removed 140 patients from within five miles of TMI to Harrisburg. Accounts of how well patients adjusted. (HP)

"Who Runs TMI?"
Sen. Hart of Colorado says the utility company has the responsibility to bring TMI under control. He also wants plants

similar in design to operate at reduced power until safety can be determined. (HP)

4/2/79 "TMI Threat of Sabotage is Ruled Out." The FBI says rumors of sabotage at TMI are baseless. (HP and LIJ)

"Jonathan Rivera and Friends Making Most of Their Plight."
How children evacuated to Hersheypark Arena are being entertained. (HP and LIJ)

"Radiation Debate Heats Over Health Dangers."
The article discusses radiation levels and possible health effects. (HP)

"Pet Owner Can't Bring Herself to Leave."
Photo story. A pet owner thinks she should leave the area, but wants to stay to care for her dog. (HP)

4/3/79 "Pres. Carter Orders Federal Inquiry to All Aspects of TMI."
Pres. Carter orders a federal inquiry into TMI to determine the causes of the accident and to make recommendations for the future of the nuclear power industry. (NYT, HP, and LIJ)

"40 Jumped the Gun on Evacuation."
People show up at reception areas in Columbia and Juniata Counties without an order being issued. Some details on Dauphin County plan. (HP)

"HHA Has Plan to Evacuate 5,000 Tenants."
Plans are outlined for residents of local housing authority units if an evacuation order is issued. (HP)

"Humane Society has Evac. Plans."
Plans are made to evacuate animals from shelters. Also, people are urged not to abandon animals as some who left the area have done. (HP)

4/3/79 "Most School Districts in Area Expect to Reopen Wednesday."
State Department of Education applies pressure to schools outside the five mile area to reopen. Reports from local school districts are presented. (HP)

"A New Governor Faces a Dramatic Dilemma."
Articles describe choices facing the Governor as the reactor cools. (LIJ)

"People Leaving Area, Job Absenteeism Up."
Despite optimistic reports from TMI, problems continued to be felt in Lancaster County as people continue to leave and the economy is hurt. (LIJ)

Editorial: "The Beginning of the End?"
The events at TMI will probably add to raising the insurance rates and the share of the damage to be covered by the owners of nuclear power plants. This may result in the costs of additional nuclear energy to be too steep for utilities and the public to bear. (LIJ)

Editorial: "Just Part of the Job."
The editorial commends Harold Denton, NRC, and other officials because no lives were lost. (LIJ)

"They're not Laughing This Week."
People are not laughing. The accident at TMI is at this stage too serious for laughter. (LIJ)

"A Calming Voice at TMI."
The article describes the calming performance of Harold Denton. (LIJ)

"TMI 3 Mile Shock Stings Market."
The stock market is lower as investors guage the impact of the accident on energy issues. (LIJ)

4/3/79 "City Preparing Evac. Plan."
Different plans are being readied depending upon the size (miles covered) of evacuation ordered. (LIJ and HP)

"Met Ed to Pay Evacuees Bills."
Met Ed will pay bills for pregnant women and preschool children who evacuated from the five mile area around TMI. (LIJ)

"Worker Describes Fight to Cool Heated Reactor: No Panic Inside Plant."
The article describes a worker's perspective on how the efforts to cool the TMI reactor are proceeding. (LIJ)

Editorial: "The Public Bubble."
The editorial states that "lies" about TMI events must not occur again. (LIJ)

Editorial: "Rumor Mill Grinds On."
Rumors continue despite decreasing chances of a catastrophe. The editorial asks who can we believe. (HP)

"Author Calls for Closing of Two N-Plants."
McKinley C. Olson, who has written about and investigated problems with N-plants since 1959, urges that TMI and Peach Bottom be closed immediately. Olson discusses radiation problems, fears, and potentials. (HP)

"Report About Governor Denied. Aide: No Kicking, Screaming."
The report is denied that Thornburgh was reluctant to issue an evaucation order on pregnant women and preschool children. An aide says his own decision was based on scientific data. (HP)

"TMI Crisis Spawns Raft of Alterations."
Temporary offices, phone lines, and other construction facilities have been set up in response to the crisis. (HP)

4/3/79

"Area Family Feels Safe in Altoona."
A Lewisberry family flees to Altoona. (HP)

"Clinton County Ready to Host 600 Evacuees."
Clinton County is prepared to host evacuees from the Harrisburg area. (HP)

"Cumberland County Pickup Areas Listed."
Contingency staging areas are announced for Dauphin and Cumberland Counties in case of evacuation. (HP)

"Newberry Official Renews Vow."
A Newberry Township official seeks to bring legal action to shut the plant forever and to have expenses resulting from the crisis paid for. (HP)

"One Hopeful TMI-Area Couple Plans Wedding in Middletown."
Reports on reactions in Middletown and Goldsboro and on one couple's plan to marry on 4/14.

"Mass Transit Figures in Exit."
Reports on gas station, bus, train, and plane business during the TMI scare and their possible role in evacuation. (HP)

"First Baby Born to Evacuee in Midst of N-Scare."
An evacuee to Hersheypark gives birth at Holy Spirit Hospital. The article describes how the center is equipped to deal with birth and pregnant women. It also reports on nursing and retirement residents evacuated outside of the area. (HP)

"'Looters will be Shot.'"
Middletown Mayor issues a shoot the looters order to the police department. Also discussed was the number of residents who left, curfew, return of some residents, and police work. (HP)

4/3/79

"Cumberland County Officials Ready for Evac."
Emergency operation center readiness in Carlisle, Cumberland County is discussed. (HP)

"Area's Second Woe: Reporters."
Reporters and news media invade Middletown. Effect on local residents and the difficulties in hearing Denton at the press conference are presented. (HP)

"Residents Going for Ready Cash."
The first story reports on the rush at banks and withdrawal of funds for travel. The second story covers bills being due and payment procedures during the emergency. (HP)

"Precautions are Continued for Residents of Midstate."
The crisis appears to be easing, but the Governor continues precautionary measures. Employee absences at state agencies is noted. (HP)

"N-Plant Plane Impact Case Delayed."
Hearings are delayed on potential danger from airplanes flying over TMI. (HP)

"Same Supplier. Other Plants with Reactors get NRC Attention."
Other plants with B&W equipment are ordered to check equipment as a precaution. (HP)

4/4/79

"H. Denton Says That Gas Bubble in Reactor is No Longer a Significant Problem."
It will take 4 years before the plant is decontaminated. Gov.'s advisory is still in effect. Denton says the bubble is no longer a significant problem at TMI. (NYT, LIJ, and HP)

4/4/79 "Schools Report Near 50% Absent."
Absenteeism was reported to be near 50% in school districts near TMI despite government assurance that there are no reasons for closing schools beyond five miles. (LIJ)

"Mood Takes U-Turn at Command Center."
Paul Leese is quoted as saying, "things are looking up." (LIJ)

Editorial: "A Brush with Doomsday."
It is feared that the American people will sink back into their former acceptance of whatever the power companies and the nuclear lobby wish to tell them. The editor feels that it will not happen here. He views the area as a miniscule part of the nation a small fraction of whose people live within the shadow of a nuclear plant. (LIJ)

"100 Raced the Clock to Get Drug to PA in Case."
Potassium Iodide is rushed to Pa. It is an antidote for radiation poisoning. (LIJ)

"Evac. Chances 'Slimmer:' County Relaxes Alert."
Dauphin County relaxed the alert status on contingency evacuation to stand by alert status. Volunteers return to jobs. (HP)

"Precaution Still Holds for Some Area Citizens."
Thornburgh announces the situation has improved; urges pregnant women and pre-school children to remain out of the area within five miles of TMI. Only schools within a five mile radius should remain closed. (HP)

"7 Days, March 28-April 4."
The article recreates the scenario of the TMI mishap. After a brief

introduction that expressed the confusion of the event, the article explains the accident through the analysis of three resulting complications: 1) How to cool the reactor; 2) How to rid the containment building of water; 3) How to rid the reactor dome of the explosive hydrogen. The steps taken by local, state, and federal government authorities are recounted. The article concludes by outlining the incidents that completed that eventful week. (MPJ)

4/4/79 "All Eyes Focus Here."
The problem of bringing about an orderly cooling of the fuel assemblies at Unit II is discussed. The extreme heat produced in the core is cited as a great concern, but the heat produced in the surrounding area as a result of losses taken by local businesses, nuclear insurers, and people worried by radiation levels is also considered to be a problem. (MPJ)

Editorial: "What Cost a Catastrophe."
The editorial begins by reminding the reader that the Middletown area has survived many situations (fires, floods), but the TMI mishap is different and harder to accept. The reason alleged for this difficulty of understanding is that the author feels nuclear power plants worry more about public image than safety. In general, the article expresses the editor's anger and frustrations charging that the nuclear industry and Met Ed were not entirely frank about the accident. The editor feels that vital information was wrongly withheld from the public. (MPJ)

"Ertel Hears Councilmen's Thoughts on Unit II Accident."
The article consists primarily of Cong. Ertel's comments regarding communication problems during the TMI accident, general

comments made by local councilmen on the value of nuclear power, and a few comments concerning Met Ed. Ertel proposed that all authority during this type of emergency be turned over to the federal level of operation, and also that a common terminology be established. These actions, Ertel felt, would help correct the communication breakdown problem and eliminate its consequent confusion. (MPJ)

"Don't Look Now Met Ed, But Your Slip is Showing."
The article comments on the confusion concerning an emergency evacuation plan for TMI area residents. Met Ed was faulted as being negligent to the needs for such a plan. Middletown Mayor Reid discusses the circumstances surrounding the hasty formulation of an evacuation plan, and the article argues that such a plan should have been completed long ago. (MPJ)

"Town Topics."
The article is one view of the TMI accident as seen from a small town newspaper reporter's standpoint. In general, the article points out some of the inconveniences suffered by area merchants due to the TMI event. (MPJ)

"And Now on the Brighter Side of Things."
Some of the more light-hearted, comic events that happened during the TMI mishap are presented. Assorted comments from people all over the world contribute to this article's attempt to see some humor in the tense situation. (MPJ)

"Disaster Aid Statement is Expected."
Dauphin County is expected to issue a disaster emergency statement. This does not mean that the situation has worsened, but opens a door for possible federal aid. (HP)

4/4/79 Editorial: "Reopening School."
The editor disagrees with the Governor. He believes that schools should be kept closed within a 20 mile radius until officials at the scene can unequivocally say the plant no longer poses a threat. (HP)

"Accident Traced to Error."
The article includes details of a *Chicago Tribune* story relative to the valve in the backup cooling system being left closed. Also, there is an account of cooling system operation. (HP)

"N-Plants to Check on Safety."
A photo story of B&W plants that were issued a federal order to run a safety check and report findings within 10 days. (HP)

"It's Open, Shut Case: Luggage Sales Are Up."
Reports from Harrisburg area stores relative to the boom in luggage sales. People are purchasing big suitcases in case evacuation is ordered and they are only allowed to take one suitcase. (HP)

"N-Peril Hurts Business."
Harrisburg restaurants and area stores report business is below normal. (HP)

"Evac. Plans at Bases Outlined."
The article details procedures for possible evacuation at New Cumberland Army Depot, and the Navy Ships Parts Control Center and Defense Depot in Mechanicsburg. (HP)

"Mother of Deaf Boy Cites Alert Problem."
The mother of a deaf boy in Middletown expressed concern that those who can't hear might be abandoned in an emergency evacuation. (HP)

4/4/79 "Newberry to Met Ed: Reimburse Us."
Newberry Township will seek reimbursement from Met Ed for all expenses incurred as a result of the TMI accident. (HP)

"Evac. Expense Payout is Outlined."
The details of federal and insurance company responsibility to individuals for personal or property damage and Met Ed's responsibility for evacuation expenses is outlined. (HP)

"Limited Liability Scored."
The Price-Anderson Act of 1957 puts a ceiling on financial liability of owners of nuclear power plants. Hampden Township Commissioners criticize the act. (HP)

"If Evac. is Ordered, You Have Right to Stay Put."
PEMA says any evacuation on a state, county, or local level would be strictly voluntary. (HP)

"Emergency Planning for Capitol is Proceeding."
Underground headquarters at the capitol are ready in case an evacuation is needed. (HP)

"Cumberland Evac. Plan: Head West."
Some plans for possible evacuation of Cumberland County are given.

"N-Data Flyers Given a Push."
Cumberland County urges the distribution of informational flyers from the Office of Emergency Preparedness relative to possible evacuation. (HP)

"Harold Denton: N-Man of Hour."
A personal background on Harold Denton with comments by his wife is provided. (HP)

4/4/79 "Lawmakers Want N-Crisis Unit."
The Pa. Congressional Delegation wants the federal government to set up a new agency to deal with future crises at nuclear power plants. (HP)

"TMI Syndrome 'A Warning.'"
The article provides comments from Mike Gray, writer of the movie *China Syndrome*, relative to the movie's message and nuclear power. (HP)

"No Lasting Effects Expected as TMI Soil is Tested."
N-physicist from Dickinson College says soil samples from nearby the TMI plant indicated that radiation released to date will probably have no lasting affects on the environment. The article also notes other soil, water, and air samplings. (HP)

"Thornburgh Looks at Long-Term Effects."
Thornburgh is relieved that the situation at TMI has improved. He wonders about long term public health, economic, and environmental effects. He also discusses state workers' absenteeism and state workers' leave policy. (HP)

"N-Books Popular."
Reports from local libraries indicate decreased patronage and popularity of books with information on nuclear power. (HP)

"Change of Scenery Perks Up Homes' Patients."
Reactions are given to the Hersheypark Arena evacuation center. (HP)

"Chambersburg Hosts 31 Evacuees Anyway."
Some York Co. residents misunderstood evacuation plans and showed up in Franklin County. Franklin County emergency plans are set forth. (HP)

4/4/79 "TMI Incident May Hamper City Bond Sale."
Financial advisers recommend that the City Council delay notices of a bond sale to build new Harristown facilities. They fear TMI will affect Harrisburg's ability to sell bonds. (HP)

"Myers Raps Operation by Met Ed."
Cumberland County Commissioner Myers blasts Met Ed's handling of the crisis. He also feels Met Ed should give over the operation of the plant to another company. (HP)

"Saxton Tested First."
Saxton, in Northern Bedford County, was the site where utility companies first researched commercial nuclear technology. The story gives the history of the Saxton facility. (HP)

"Exodus Plan Imperiled."
Dauphin County CD officials feel that the reopening of schools and returning to normal will imperil their contingency evacuation plans. (HP)

"Reporter at Large."
The story details how Harrisburgers have coped with disasters and problems over time. (HP)

"Most Area Classes Resume Wednesday."
The article includes a listing of local area school and college class resumption schedules. (HP)

"Reopen TMI?"
The article questions whether it is worth the risk of endangering our lives, our land, and our future in order to extract 13% of this country's electrical power from nuclear plants. The argument in the article is based on the unseen, unknown potential hazards of nuclear fission versus the known hazards of coal use for electricity. (MPJ)

4/4/79 "The Press and Journal Asks..."
Area residents' responses to the reporter's question, "What precautions are you taking to off-set the nuclear leak at TMI?" are presented. (MPJ)

"First-Ever Visit by U.S. President Provides Encouragement for All."
A review of Carter's activities during his TMI tour. A few of his comments on the situation are given. (MPJ)

"Anxious Residents Reflect Opinions Offered by Mayor."
A review of a few of the comments made by Middletown residents and their reactions to the events. The article concludes with residents' comments on their personal solutions to the stress situation. (MPJ)

4/5/79 "Nuclear Jitters."
The story outlines different people's emotional reactions to the accident. (LIJ)

"Crisis Declared Over at TMI as Cold Shut Down Continues."
"The threat of an immediate catastrophe is over," said Gov. Thornburgh. (LIJ)

"NRC Says Serious Operating Errors Caused Mishap at TMI."
NRC officials state that improper closing of two key valves caused the mishap at TMI. No shutdown of eight other similar reactors in the US is ordered. (LIJ)

"Hershey Foods Delays Use of Some of its Milk."
Hershey Foods temporarily discontinues the use of milk shipped to local plants. Other precautions are taken by the company. Some discussion on Iodine 131 is included. (HP)

4/5/79 Editorial: "Heroes in a Crisis."
The editorial praises the calmness of Harold Denton, the reassurance he brought to the area, and compares him to a father figure. (HP)

"Heavy Political Fallout from TMI."
The effects of TMI on proposed nuclear legislation and the building of future nuclear plants are discussed. It calls for probes at the federal level. The reactions of various politicians and federal officials to TMI are provided. (HP)

"To Our Friends Beyond the Radius, Hi!"
An open letter by a local reporter making light of experiences during the week following the accident. The letter is directed to people outside the area. (HP)

"County Seeks U.S. Aid."
Dauphin County Commissioners declare a disaster emergency. This allows them to apply for possible federal aid. They also ask Pres. Carter to reconsider the proposed closing of Fort Indiantown Gap and the New Cumberland Army Depot in light of the recent crisis. (HP)

"Mrs. Thornburgh Visits Group."
The Governor's wife visits Hersheypark Arena, an evacuation site. Activities there are discussed. (HP)

"Middletown Mulls Resolution."
Middletown Council ponders drawing up a resolution relative to the state of emergency. They feel the resolution will enhance chances of receiving federal aid. (HP)

"Officials in a Tiff on Buses."
A CD official in Londonderry Township and the Lower Dauphin School Board

Authority clash as to who has jurisdiction over school buses that might be used for evacuation. The issues are the fear of stealing buses with keys versus having vehicles ready for evacuation use. (HP)

4/5/79 "York County Misgivings Still Evident." A report on fears, absenteeism, and reactions of York County residents relative to TMI. (HP)

"Some Reach for Humor." Dickinson College students and their T-shirt slogans relative to TMI are highlighted. (HP)

"Curfews." Curfews are lifted by local communities. (HP)

"Posterity Will Reap Crisis Lore." Dickinson College sociology students call for TMI crisis folklore. (HP)

"TMI Suit is Filed, Withdrawn." A couple filed suit against Met Ed and B&W to shut TMI and give compensation to everyone living within a five mile radius. (HP)

"Return to Normalcy Urged." A state CD director calls for a return to normalcy from preoccupation with nuclear energy. The article comments on possible problems for the business community. (HP)

"Most Schools Report a Nearly Normal Day." Attendance reports and activities at area schools are outlined. (HP)

"N-Foes Set Rally for Sunday." TMIA will hold a rally at the State Capitol. (HP)

4/5/79 PUC Aide Ends Evac. He Hadn't Planned to Take."
A report is given of one man's evacuation and what it meant to him. (HP)

"68% in Poll Say Danger of Radiation is Greater."
A CBS News poll finds that 68% of Philadelphians believe that the accident was worse than stated and only 16% say they believe officials were being honest. (NYT)

"Errors Confirmed in N-Incident."
NRC officials give details of two human errors and one mechanical error that led to the TMI problem. Details on tentative cooling down of the reactor are given. (HP)

"Zeller Suspects Sabotage."
State Senator Joseph Zeller contends that these things don't just happen and suspects sabotage at TMI. Zeller doesn't want the US to become a second rate nuclear power. He also criticizes Schlesinger's handling of facts. (HP)

"Radiation's Long-Range Effects Unclear."
Senators Kennedy and Schweiker question witnesses including Joseph Califano, Secretary of HEW, concerning the impact of TMI radiation on health. (HP)

4/6/79 "Middletown Electric Bill Won't Jump."
Met Ed increases will follow as a result of the accident, but Middletown rates will not increase. Middletown has been protected from rate increases since a 1906 agreement with Met Ed. (HP)

"Area Evacuees 'Doing Fine' at New Site."
The article reports on individuals evaucated to Hersheypark Arena as well as persons evacuated from area nursing homes. (HP)

4/6/79 "N-Probe Plan Unveiled."
Pres. Carter announces a plan for an independent panel to probe TMI. (HP, LIJ, and NYT)

"TMI 'Rush Job' Denied."
The NRC condemns Ralph Nader's allegation that TMI was rushed into operation. The differentiation between declaring the plant commercial as opposed to licensing is outlined. Comments from Harold Denton relative to the cooldown are given. (HP, LIJ, and NYT)

"Brief Devotion Eases a Crisis."
Monsignor William Keeler holds a brief devotional service for employees and volunteers at Dauphin County's Office of Emergency Preparedness. Also included are comments by Dauphin County officials and their feelings about closing TMI. (HP)

"Radiologist Sees Limited Danger."
Radiological emissions at TMI are too small to cause illness according to a Hershey radiologist. (HP)

"Radiation May Cause One or Two Cancer Deaths."
The DER releases a report on projected cancer levels and exposure to radiation. One or two cancer deaths are possible from TMI. The average personal exposure is far below national safety guidelines. (HP)

"City to Tell its N-Story."
Harrisburg City will provide information on TMI in its prospectus for a new bond issue. TMI is believed to have financial implications that could hamper the city's ability to float bonds. (HP)

"It Could be Years or Decades Before Reactor is Repaired."
The article details what clean up of the reactor may involve in terms of costs and time. (HP)

4/6/79

"Insurance Payout on TMI."
Evacuees from TMI's accident received more money from two insurance pools covering the nuclear industry than victims of any other nuclear related accident. Insurance claims and the nuclear insurance industry are also discussed. (HP and LIJ)

"TMI Consumer Cost: $7.50 a Month."
A power company attorney estimates the cost to the consumer will be $7.50 per month due to the TMI accident. Legislation is pending to make passing of costs on to the consumer unlawful. (HP)

Editorial: "Area Recovery: Feds Should Provide Assistance."
The editor discusses the psychological and economic impacts of the crisis. He calls for Gov. Thornburgh to declare the region an economic hardship and for federal legislators to introduce legislation needed to meet area needs. The editor concludes that South Central Pa. has overcome hardships in the past and will rebound again. (HP)

"Evacuated Boy Killed in Crash."
A two year old boy, whose family had evacuated Middletown, was killed in a car crash. His family had been staying in Hershey. (HP)

"Robot Ready to Help Out at N-Plant."
Herman, the Mechanical Manipulator with the ability to resist radioactivity, has been flown from Oak Ridge, Tenn. to aid in determining the extent of core damage at Unit II. The story gives details of how Herman works. (HP)

"Cumberland Acts to Recover TMI Emergency Expenses."
Cumberland County declares a state of emergency retroactive to the TMI incident. They hope to gain federal assistance for bills incurred during the

emergency. The county plans to submit bills for overtime payment to county employees to Met Ed and its insurance carriers. (HP)

4/6/79 "New Cumberland Seeks Payment."
New Cumberland Borough is seeking reimbursement from Met Ed for expenses incurred during the TMI emergency. (HP)

"Crew Worked 40 Straight Days Before Accident."
The article reports that maintenance crews at TMI worked for more than five weeks without a day off prior to the accident. Many were ten hour days. The work schedule in relation to valve error is discussed. (HP)

"Citizen Comment Invited by Ertel on Emergency."
Cong. Ertel will hold a town meeting in Middletown to gather citizen reaction to the incident. (HP)

"TMI Crisis Prompts Company to Move Trade Show from City."
American Hardware Supply Company announces the movement of their week long trade show from Harrisburg to Memphis. The show would bring $1 million to the local economy. (HP)

"City Seeks to Recoup TMI-Related Expenses."
A list of expenses incurred by Harrisburg during the TMI crisis and some preliminary plans for recouping them are discussed. (HP)

"Class Action Suit Filed Against TMI."
A class action suit was filed in Dauphin County Court on behalf of business and individuals within a 20 mile radius of TMI. Allegations in the suit are detailed. (HP)

4/6/79

"PUC Will Probe Accident Impact."
The PUC launches an investigation into the financial impact that TMI will have Met Ed customers. The probe refers to proposed rate hikes and raises questions relative to who will pay the costs of the TMI accident. (HP)

"Among Nuke Plant Workers, Faith in Three Mile Unshaken."
Judging from a number of interviews with plant workers, the reporter concluded that their faith in TMI has not been shaken. (LIJ)

"Pregnant Women, Tot Stay-Away Continues."
Harold Denton stated that the evacuation of pregnant women and small children from the vicinity of TMI will continue until the flooded floor of the auxiliary building at the nuclear plant can be substantially decontaminated. (LIJ)

"Calls Show Hundreds of Pregnant Women Worried."
Local abortion clinics and gynecologists say that they are handling hundreds of calls from worried pregnant women living near the crippled TMI nuclear plant. (LIJ)

"Be Available, Evacuation Plan Personnel Urged."
LEMA personnel were urged to remain on alert until the Governor orders that an alert is no longer needed. (LIJ)

"Nuke Plant Inspectors Bill Eyed."
A trio of Central Pa. congressmen have introduced legislation which would require all nuclear power plants in the country to have resident NRC inspectors. Cong. Walker also announced that the House Committee of Science and Technology will hold hearings on the safety of nuclear reactors on 5/22-5/24. (LIJ)

4/6/79 "Anti-Nuclear Group Plans 'Concern' Rally Saturday."
The first public rally of the newly formed antinuclear group in Lancaster, called SVA, was announced. Rev. R. Vieth and J. Kohr, Attorney, will be speakers. Representatives from the Keystone Alliance (a Delaware Valley antinuclear group) will also be present. (LIJ)

4/7/79 "Two 'Cautions' But Pro-Nuclear Congressmen."
Congressmen Ertel and Walker indicate a pronuclear stand with caution. Both advocate improved communication to prevent information breakdown in the future. The article also contains comments from Cong. Goodling concerning reports put out by politicians during the crisis. (HP)

Editorial: "N-Role: We Need a Choice of Scenarios."
The editorial discusses nuclear energy in light of the overall energy crisis. It calls for studies and comparison of nuclear energy versus other forms, and questions whether the country can proceed with nuclear energy or not. (HP)

"Scranton See Economic Strains in TMI Publicity."
The Lt. Gov. expects the area to suffer as a result of publicity connected with TMI. Scranton comments on the roles of Gov. Thornburgh and Harold Denton during the crisis. (HP, LIJ, and NYT)

"TMI Shadow on Bonds. Advisers Differ."
Financial advisers warn Harrisburg that they may not be able to float city bonds as a result of the crisis. A city business administrator differs with the advice of the counsellors, but states that he has not lost confidence in them. (HP)

4/7/79 "10-20 Hour Exit is Guess for City."
Harrisburg's Police Chief believes that the evacuation plan developed for the city would take ten to twenty hours to implement. Some details are provided on the plan and what an evacuation order would have meant. (HP)

"Impact Pleas in Accident at TMI Filed."
Consumer advocate, Mark Widoff, files actions with the PUC to soften the economic impact of TMI. His pleas involve a pending Met Ed price hike, a rate increase given to Penelec in 1/79, and not automatically passing on the costs of the TMI disaster to the consumer. (HP)

"Crisis Spawns Trailer City of 1,000 Souls."
Trailer City, a picnic grove and cornfield, becomes home for individuals working on the TMI crisis. It has more than 26 mobile homes, a newspaper, mail delivery, and a helipad. It is located behind Met Ed's Observation Center. (HP)

"Shirt Tale."
A photo story of modeling a TMI T-shirt. (HP)

"TMI Panel Chairman Choice Due."
The article discusses Pres. Carter's plans to form a commission to investigate the TMI problem. (HP)

"Thornburgh Doubts 3 Mile Should Reopen."
Thornburgh expresses doubts about the future of nuclear power and whether TMI should ever reopen. The Governor details actions he is calling for, such as federal aid, investigation, and a review of the role of nuclear power in Pa. (HP)

4/7/79 "IRS Extends Tax Deadlines of Evacuees."
An automatic thirty day tax filing extension is granted to evacuees in certain Pa. counties. (HP)

"Radioactive Gas Leak is Plugged."
A leak is discovered in pipes involved in pumping radioactive gas back into Unit II. The article contains comments by Harold Denton concerning evacuees returning to the area, and cooling down the reactor. (HP)

"Experts Overcome 2 Problems on 'Quiet Day' at TMI."
Two equipment failures, a pump and piping system, were overcome as experts continue their efforts to bring the crippled TMI nuclear power plant to a cold and stable shutdown. (LIJ)

"Hess: Met Ed Should Pay Off."
State Sen. R. Hess requested that the PUC prevent Met Ed from having its customers pay the cost of the nuclear accident at TMI. (LIJ)

4/8/79 "Pennsylvanians Given Assurances."
The Gov. and Denton say the threat to the people is lessening as the cooling down continues. (NYT and HP)

4/9/79 "Lowering of Pressure in Reactor Coolant System Continues."
For the second day at TMI, a lowering of pressure in the reactor coolant system occurred. One thousand protestors at the State Capitol demand that TMI remain permanently closed. (NYT, HP, and LIJ)

"NRC Faces Test of Credibility."
The article discusses the impact of TMI on the NRC's future. The agency will undergo a challenge of credibility. The NRC has been under fire and is sometimes viewed as advocating nuclear power at public expense. (HP)

4/9/79

"NRC Chairman Visits Command Post."
Joseph Hendrie visits the command post at TMI. (HP)

"Congress to Open TMI Probe."
The details are revealed of plans for Senate and House hearings related to TMI. (HP)

Editorial: "Energy Future: State Lacks Direction."
The editorial states that what is needed more than anything in the wake of TMI is an investigation of the future of energy in Pa. It asks who is in charge; contends no one. He asks why the state has gone heavily into nuclear power. It advocates moving in the direction of coal. Finally, the editorial discusses energy related issues such as the use of solar power and public transportation. (HP)

"TMI Blame Laid by Schlesinger."
James Schlesinger believes the blame for TMI will fall on people and management, not equipment. He feels that the nuclear power industry is a necessity. (HP and LIJ)

"Cold Shutdown Process Begins at TMI."
The arrival of supplies for beginning the cooldown procedure at TMI is highlighted. (HP)

"Gov. Expected to Declare TMI Area Safe for Pregnant Women, Tots."
Because engineers at TMI have succeeded in controlling the two primary sources of low level radiation that were leaking from the plant, the Governor's advisory for the five mile area may be lifted. (LIJ)

Editorial: "Never, Never Again."
The editorial contends that nuclear power plants are not safe. (LIJ)

4/9/79 "N-Accident; TMI Plant."
This article explains errors, equipment failures, potential hazards, elevated radiation levels, and plant problems following the accident. It also includes a summary of cleanup techniques along with a rundown of the effects of the accident on antinuclear organizations. Finally, the report explains the negative effects inflicted on the nation's energy crunch. It states that the plant's nearly disasterous accident did nothing to boost people's acceptance of nuclear energy as an alternative power source to coal and oil. (NW)

4/10/79 "Poll Shows Sharp Rise Since '77 in Opposition to Nuclear Plants."
A NYT/CBS News poll taken after the accident finds a sharp falling off of public support for nuclear power; opposition strongest among women. (NYT)

"Faculty Asks Moratorium, Phase Out of N-Plants."
Urging protests and energy conservation, the faculty of the Lancaster Theological Seminary urged the phasing out of nuclear power plants in the US. (LIJ)

"Columbia Will Submit Petition to Thornburgh."
Columbia Borough Council passed a resolution urging the Governor and Met Ed to establish better ways of releasing information in an emergency situation regarding the nuclear power plant at TMI. (LIJ)

"Denton Says Crisis Ended; Evacuees Told to Go Home."
Denton said that the crisis at TMI is over with regard to the status of the core. The Governor lifts the advisory for the five mile radius and orders the reopening of all schools. (LIJ, NYT, and HP; also 4/11/79 MPJ)

4/10/79 Editorial: "Who Pays? Don't Let the Company Off the Hook."
The editorial advocates that customers should not be the ones to pay the cost of TMI, either in terms of fairness or lessons to be learned by utilities regarding the necessity of safe operations. (HP)

"Consumer Aide Claims Meddling."
Mark Widoff charges that the State Attorney General is interfering with his office following the TMI accident. (HP)

"Rehearing of Met Ed Rate Sought."
The PUC trial staff calls for a stay and rehearing on the rate given to Met Ed. The request is based on Unit II being out of commercial operation. (HP)

"Don't Force Public to Pay, Bill Demands."
Legislation to prevent consumers from footing the bill on TMI is introduced in the State House. (HP)

"City Credit Review 'Welcomed.'"
Moody's Investors Service plans a review of Harrisburg's credit rating in connection with the proposed bond issue. Mayor Doutrich claims this gives the city an opportunity to present an improved financial picture. (HP)

"Free Tests of Radioactivity Will Be Given."
The Pa. DER will conduct tests to see if area residents have increased radioactivity in their bodies. Details of test eligibility and sites are provided. (HP; also 4/11/79 MPJ)

"Goldsboro Will Seek Damages."
The Goldsboro Borough Council criticizes York County's emergency procedures and authorizes claims against Met Ed for costs incurred during the crisis. (HP)

4/10/79 "Thornburgh Lifts N-Accident Alert."
Gov. Thornburgh lifts the eight day precautionary directions and urges a return to normalcy. He announces Lt. Gov. Scranton will chair a Central Pa. Recovery Committee. (HP)

"City Crime Rose 29% in 5 Days."
Harrisburg's crime rate involving burglary, robbery, and auto thefts rose 29% in a five day period when many residents had fled the city. (HP)

"Evacuees Elated."
Local people, expressing relief, return to the area. (HP)

"Meeting of the Minds Fights Vagaries of TMI."
Several hundred nuclear scientists and experts gathered near TMI during the crisis. The article describes their daily routines, coping, and advice they provided. (HP)

"Counties Seek 'United Voice' in N-Planning."
Cumberland County Commissioners call for all area county officials to work together to resolve problems of TMI on a regional basis. (HP)

4/11/79 "Area Leaders Act to Lighten TMI-Born Burdens."
Business, government, labor, industry, and civic leaders meet and decide to fight bad publicity for the area in the wake of TMI. A task force is formed. (HP)

Editorial: "State Strategy. Consumer Advocate Undermined."
The editor states it was wrong for Thornburgh's Administration to hand coordinated control of TMI related matters over to the Attorney General. The editor fears a loss of independence. (HP)

4/11/79

"Radiation Fear is Irrational."
The fear of people in the TMI area is termed irrational. The article states it threatens the future of the nuclear power industry. It discusses the history of nuclear power and antinuclear power groups. (HP)

"TMI Residents 'Clean.'"
A report on body scans administered to people near TMI. No evidence of accumulated radioactive isotopes is found. (HP)

"City Council Wants TMI Shut Down."
The Harrisburg City Council calls for a shutdown of TMI with resumption of operations contingent on fail-safe methods of insuring safety. It calls for Met Ed to bear the costs for monitoring the health of persons within a 20 mile radius for 20 years. (HP)

"Panel to Subpeona NRC Secret-Talks Logs."
A House subcommittee voted to subpoena the minutes of nine secret NRC meetings held to discuss TMI between 3/30 and 4/9. They believe the NRC had no right to meet secretly. (HP)

"Township Knocks York County CD."
Fairview Township supervisors criticize York County CD. The township had developed its own plan and alleges that pressure was put on them to follow the county plan. (HP)

"Cumberland Acts to Assess TMI Impact."
Cumberland County Commissioners call for a task force for area recovery. They feel psychological fallout has occurred. The article lists some economic and social effects resulting from TMI. (HP)

"5 Firms Seek $560 Million in N-Suit."
A class action suit was filed in a US Middle District Court alleging that the

TMI accident resulted from negligence. The suit will represent individuals, proprietorships, partnerships, corporations, and other business and professional entities within a 20-mile radius of TMI which suffered economic harms linked to the TMI accident. (HP)

4/11/79 "TMI Sleuths in Bind During Early Days." The article reports on the testimony of NRC's Chairman before a Senate Subcommittee on TMI. It alleges that efforts to cooldown the reactor were impeded by inadequate monitoring, incomplete information, and poor communication. (HP)

"Rems, Rods, Roentgens."
The properties and characteristics of radiation are explained. The three terms mentioned in the article's title are specifically defined. Radiation's effects on humans are reviewed in terms of dosage levels. (MPJ)

"Cong. Ertel Attempts to Answer Town's Questions About TMI."
A report on a town meeting conducted by Cong. Ertel. The article centers on the citizens' complaints regarding the payment for the Unit II accident which would be represented in higher future utility bills. The main concern of the meeting appears to have been the problem of who should pay for such an accident. (MPJ)

"Met Ed Customers; Brace Yourselves."
The article warns the reader that the public can expect to pay higher utility bills in order to cover the Unit II repair bill. (MPJ)

"Bill to Stop Met Ed From Pushing Costs Onto Consumers Introduced."
The filing of legislation that would prohibit a Met Ed utility rate increase (reflecting costs of the TMI accident unless properly authorized) is reported. (MPJ)

4/11/79 "Threat of Radiation in Water Dispelled." TMI residents may be assured that their drinking water is not contaminated with radiation. (MPJ)

"Evacuate Before You Evaporate." Amusing comments on TMI from local folk as well as outsiders. (MPJ)

"Japanese Reporter Relates TMI Impact." A Japanese reporter provides his observations on nuclear power use in Japan. The Japanese reporter felt that the TMI mishap had little impact on Japan's nuclear industry. (MPJ)

"The Press and Journal Asks..." This column contains responses from people who were asked to give their opinions concerning the imposing of repair costs for TMI on the customers of Met Ed. (MPJ)

"School Legislation in Hopper Due to TMI Mishap." Legislation which would not require graduating students to return after graduation to make up the days lost due to the TMI events is discussed. State Rep. Rudy Dininni, who introduced the bill, felt that students' post-graduation plans should not have been affected by something over which they had no control. (MPJ)

"Public Safety is Given Priority in TMI Cooling." The public is assured that the long term cooling procedure proposed for use in the Unit II reactor was carefully considered with regard to the public's safety. Met Ed commented that the company would defer to the NRC as a sole source for information on TMI since the need for one official source was imminent. (MPJ)

4/11/79 "IRS Extends Tax Deadline for Evacuees." Local residents learn that they are entitled to a 30 day extension on the deadline for filing individual tax returns as ordered by the IRS. (MPJ)

"Catholic Editor Calls TMI Accident A Moral Issue."
Father Thomas R. Haney argues that the nuclear system is geared to profits rather than to people. (MPJ)

"Marketing Board Says Milk is Safe to Drink."
The public is advised that TMI area milk is safe to drink. Test results are given to support the claim. (MPJ)

"Radiation Tests Show No Abnormal Readings."
Scores of people who live near the crippled TMI plant received free radiation tests and were told that the computer checks confirmed that the levels in their bodies were normal. Arrangements for tests were made by the NRC and carried out by the Hegelson Nuclear Services Co. of the San Francisco area. (LIJ)

"Class Action Suit is Filed Against TMI."
Five persons have filed a class action suit against Met Ed asking for compensation due to gross negligence on the part of the owners and operators of TMI. (LIJ)

"Pres. Carter Says US Cannot Abandon Nuclear Power in Forseeable Future."
Residents near TMI receive free radiation tests which show normal levels. Government records show constant problems at TMI from 4/78 to 1/79. (NYT)

4/12/79 "Pres. Carter Names Members of Commission with John G. Kemeny of Dartmouth as its Head."
Pres. Carter names John Kemeny of

Dartmouth, a computer specialist, to head the Presidential Commission to investigate the TMI accident. Names and affiliations of the members of the Commission are given. (NYT, HP, and LIJ)

4/12/79 "City Borrowing Power Not Affected, After All."
Word from Moody's Investors indicates that the incident at TMI should not have a material effect on the rating of Harrisburg City Bonds. The bond situation is recounted. (HP)

"Hearing on TMI Due Vote."
HRPD will vote on holding public hearings on emotional and physical health concerns of area residents in connection with TMI. HRPD is the Health System Agency for South Central Pa. (HP)

"Perry Asks State for Reserve Gas."
Perry County Commissioners asked the State to help avert a shortage of gasoline that threatens them in the wake of TMI. They ask for the use of part of an emergency allotment that is made available to the State by major oil suppliers to handle emergency needs. (HP)

Editorial: "Road to Recovery. The Area's Main Asset is People."
The editor states that the determination of Central Pennsylvanians in the face of TMI and its impact is an asset. It will demonstrate that the area is still a good place to live, work, and visit. (HP)

"N-Relic a Tribute to Pride."
The article provides details of an atomic power plant in Piqua, Ohio. The plant is no longer in operation. It's viewed as a source of pride to the residents. This view is contrasted to feelings about TMI. (HP)

4/12/79 "TMI Sparks Safety Call."
This article provides an account of safety checks and actions at other N-plants as a result of the TMI accident. (HP)

"SC Bars N-Waste."
SC Gov. Richard Riley refuses to allow trucks bearing TMI waste to enter a dump site there. The waste is thought to be above low level. (HP)

"Cold Shutdown Plans Studied."
Possible cooldown procedures are discussed. (HP)

"Gradual Cooldown Set."
The NRC authorizes a gradual cooldown at Unit II. The article discusses cooldown procedures and also raises questions on the controversy of who should pay for cleanup. (HP and LIJ)

Editorial: "TMI Inquiry. Given an Impossible Task."
The editorial discusses the Commission formed by Pres. Carter to investigate TMI. It calls for the Commission to follow an independent path, and to resist any official policy. (HP)

"Schlesinger Sees Lessons in TMI."
Schlesinger still believes in the future of nuclear power. He lists the lessons learned as the need for proper procedures, proper supervision, extended training, and quality assurance in the operation of nuclear power plants. (LIJ)

"80% of Women See TMI Crisis as Very Serious."
A primary finding of the telephone survey conducted by Dr. Don Kraybill of Elizabethtown College is that women are more negative about TMI than men. (LIJ)

4/12/79 "Walker Seeks Nuclear Power Future Views."
Cong. Walker is undertaking a random survey of the 16th Congressional District residents to assess their opinions of the future of nuclear power, in light of the recent TMI mishap. Cong. Walker is a member of the House Committee on Science and Technology. (LIJ)

4/13/79 "Nuclear Officials Feared a Disaster in First Days After Mishap."
Transcripts of a closed NRC meeting on the TMI accident show that officials were very concerned for a while about a possible disaster. (NYT; also 4/14/79 HP)

"NRC Issues Warnings."
Warnings to the operators of 34 atomic plants against the mistakes and breakdowns that caused the accident at TMI are detailed. (NYT, HP, and LIJ)

"Who's in Charge Here? Governor, Ertel Finds."
Cong. Ertel questions his role in decision making during the TMI crisis. He found that he was rebuffed in attempts to get close to the media and those in charge. The Governor's press secretary details responsibilities of the NRC and of the Commonwealth. There was no role for Cong. Ertel. (LIJ)

"State and Federal Agencies Find no Radioactivity."
A report of government officials indicates that no traces of radioactivity were found in ground and surface water collected from the vicinity of TMI. The State DER, USEPA, and the US Geological Survey participated in this study. (LIJ and HP)

4/13/79 "Met Ed Workers Who Evacuated Will Get Paid."
Met Ed reversed an earlier position and will pay female employees who are pregnant or mothers of preschoolers who resided in the five mile radius and evacuated. (LIJ and HP)

"Western PA, Polled on TMI."
A survey of 600 persons by KDKA TV of Pittsburgh revealed that a majority felt that the situation at TMI was not handled well. (LIJ)

Editorial: "The Ostrich Syndrome."
The editorial states that the issue at the moment is not the future of all nuclear energy plants ... it is the future of TMI. Met Ed had the opportunity to operate a nuclear plant safely. The editorial alleges that they did not do this. Therefore, Met Ed does not deserve a second chance. (LIJ)

"Rate Act Delayed."
A $49 million rate hike for Met Ed has been postponed by the PUC for one week. The PUC will also deal with requests that consumers be spared the costs of cleanup. (HP)

"Middletown Mother Named to Study TMI Accident."
The photo story of Anne Trunk and family. She was named by Pres. Carter to the Commission to study TMI. (HP)

"TMI Claims Estimated."
Insurers of Unit II estimate that damage to the facility will probably exceed $140 million and liability claims will exceed another million. Also discussed are costs, distribution of claims to evacuees, procedures for gaining reimbursement, and how nuclear insurance liability claims are paid out. (HP)

4/13/79 "$560 Million Class-Action. TMI Suit Filed Here."
A class action suit on behalf of residents within 20 miles of TMI was filed in US Middle District Court. The suit was filed by the same attorneys who earlier filed a class action suit on behalf of area businesses. The suit alleges neglect by builders and owners of TMI. (HP)

"NRC Says Little or No Uranium Fuel Melted."
The NRC says a preliminary analysis of the coolant that circulates around the Unit II reactor indicates that little or no uranium fuel melted in the accident. It also reports on the failure of a measuring channel in the reactor's pressurizer. (HP)

"TMI Health-Related Hearings Planned."
The Board of Directors of HRPD voted to hold public hearings on emotional and physical health concerns of individuals living in the TMI area. (HP)

4/14/79 "Advisory on Area Herd Lifted."
The Pa. Secretary of Agriculture cancelled an advisory to area farmers to keep dairy animals on stored food rather than pasture. Tests of iodine samples on grass indicate no concern. (HP)

"'Recovery' Endeavor Backed."
A Harrisburg area recovery task force, called 'Forward,' receives endorsement from Lt. Gov. Scranton. The group wants to insure area residents that their health is safe and to market an improved image for the area. (HP)

"W. Shore 'Body Scan' Number Set."
West Shore residents living within three miles of TMI can call a special number for an appointment for a body scan to test internal radiation levels. (HP)

4/14/79 "TMI Foes to Fight Met Ed Rate."
The Newberry Township TMI Steering Committee plans to petition the PUC to reconsider a rate increase given to Met Ed. They want the utility to be prevented from passing cleanup costs to the consumer. The group is also calling for a permanent shutdown. (HP)

"Doctors Say Radiation No Problem to Unborn."
The article presents comments by the American College of Obstetricians and Gynecologists and the American College of Radiologists as to the dangers of radiation at TMI for pregnant women. (HP)

"T.V. Station Airs Some Late News."
A Philadelphia television station accidentally interrupted a soap opera to broadcast a two week old report warning of a possible meltdown at TMI. Some area residents clog telephone lines seeking reassurance. (HP)

"Crisis Ripples Analyzed."
High school students are asked if there should be a discouraging outlook for the TMI area. They were also asked how they would convince others that it is a safe place. (HP)

"Final Phase of Cool Down is Underway."
Federal officials are pleased with the progress in cooling down the reactor and believe they are now in the final phase of this process. (LIJ; also 4/15/79 NYT)

"Water is OK in Area Near Nuke Plant."
Further testing by three federal agencies indicates that the water is ok. (LIJ)

4/16/79 "No Harm is Seen in TMI Radiation."
Yale University radiation experts say that the amount of radiation released

in the TMI accident did not endanger the population in the nearby area. They did agree with the Governor's decision to evacuate. (HP and LIJ)

4/16/79 "Fuel Wasn't Problem During TMI Crisis." Emergency supplies of gasoline for evacuation purposes were brought into the TMI area during the emergency situation. (HP)

"Utilities Unprepared, NRC Says." NRC Commissioner, John F. Ahearne, feels that other utility companies operating nuclear reactors around the country would not have been any better prepared for a crisis than Met Ed. Information on tapes of closed door hearings on the crisis indicate that the NRC had distrust of Met Ed's technical capabilities at the site. (HP and LIJ)

Editorial: "NRC Tapes: A Callous, Unacceptable Attitude." The editorial states that the minutes of the NRC during the crisis indicate that no one in charge knew what was going on during that period. It charges that officials knew the situation was serious, but were preoccupied with issuing statements of reassurance. An attitude change on the part of officials from one of gambling with lives to protecting lives is called for. (HP)

"O'Leary Now Sings N-Song." Jack Anderson, national columnist, charges that O'Leary who came to the DOE from the AEC did an about face from being a regulator to being a paper tiger. Anderson claims the Carter administration is relaxing standards on nuclear plant licensing rather than tightening them. (HP)

"Iodine Traces Found Near Plant." Traces of radioactive Iodine 131 have been found in samples taken in a five mile radius of TMI. Levels of iodine have been low enough that federal officials discount danger. The article also discusses efforts at cooling the reactor. (HP)

4/16/79 "Reactor's Core Damaged Extensively Early in Crisis."
A NRC official reported that the reactor core suffered grave damage in the early stages of the crisis. (LIJ)

"Nuclear Incident Poses No Real Health Problems."
A report of the CDC indicated that few health problems will be caused by the TMI accident. Uncertainty exists about the impact of low level radiation. (LIJ)

"Cooling Off the Crisis; TMI Plant."
This article is an account of the causes, threats, long term problems at TMI, and the accident itself. The violations that TMI's operators committed against the NRC's regulations are explained. It describes how equipment failure, design flaws, and human errors all contributed to the TMI incident. The danger of the developing hydrogen bubble in the reactor is explained in easily understood terms. The cleanup of the left-over Krypton and highly radioactive water is explained, and future plans, potential problems, and solutions involved in the cleanup are offered. The article concludes with a general summary of the evacuation dilemma, the estimated radiation dosage received by the population varying distances from TMI, and the overall feeling of thankfulness felt since the accident was less serious than it could have been. (NW)

"Beyond the *China Syndrome;* TMI Accident."
This article reviews the movie thriller, *China Syndrome*, in terms of its

similarities to and differences from the TMI accident, which occurred only twelve days after the film's release. (NW)

4/16/79 "Covering TMI; Press Coverage." This article describes methods used by many reporters to cover the TMI incident. Overall the job done by reporters is judged to have been thorough and responsible taking into consideration the technical and often contradictory information given to the press. Some scattered examples of sensationalism are given. The report also shows how some new organizations equipped their reporters with safety and gauging devices in order to provide them with some protection from excessive radiation. Finally, the article points out that despite the press' efforts to report accurately the mishap, reporters were still accused of inadequate accounts. Specifically, the press is said to have shown restraint with such a serious topic in order to prevent panic, which unfairly led to accusations of a coverup instead of eliciting praise for the use of discretion. (NW)

"As I Was Saying." George Will, national columnist, comments on what he seems to think is unjust negative criticism of nuclear power as an energy source as a result of the TMI mishap. Will argues that new technologies often fail and their revision and improvement are results of trial and error. Will explains that nuclear power has a remarkable safety record up to the TMI accident, and the hazards from smoking, traffic situations, and coal mining and burning should be more carefully evaluated. (NW)

"Nuclear Power on the Ropes." Nuclear power in general is discussed. Opinions of political figures, planned changes in safety operations in plants, waste disposal problems, financial woes to customers, and the necessity of nuclear energy in view of coal's drawbacks are all mentioned. Relative to TMI, this article touches on the accident's effects on foreign views of nuclear power, Met Ed's great financial woes, and TMI's potential cleanup techniques. (NW)

4/17/79 "DER Will Take More Milk Samples From Farmers."
The DER announces milk samples will be taken from farms near the plant. Iodine emissions have been exceeding daily permissible rates. However, a spokesman feels that the milk sampling will be an academic exercise. The article also refers to progress toward a cold shutdown of Unit II and announces that the NRC's Division of Inspection and Enforcement has begun a formal investigation of the accident. (HP and LIJ)

"N-Plant Shutdown Pushed."
Middletown residents urged Borough Council to seek a permanent shutdown of TMI. A physician, Dr. John Barnoski, states that there is still danger. He is seeing patients with stress related complaints. Both Middletown and Steelton Borough Councils have adopted state of emergency resolutions. (HP)

4/18/79 Editorial: "A Job Well Done."
Favorable praise is given to Harold Denton for his work at the nuclear plant at TMI from 3/30/79 to 4/17/79. (LIJ)

"Denton Goes Home After Taming TMI."
Denton left TMI to return to Washington as radiation iodine leaks have been remedied and the reactor itself is approaching cold shutdown. (LIJ, HP, and NYT)

"Met Ed, Public Argues Costs."
At a PUC hearing, Met Ed argues for a rate increase to be implemented. The

increase was delayed due to the TMI accident. Met Ed says it will go bankrupt if the increase is not implemented. M. Widoff opposes the increase saying insurance should pay for increased costs. (LIJ and HP)

4/18/79 "2 Rigs Carry TMI Waste West." Tractor trailers are carrying TMI waste to Richland, Wash. The article discusses the nature of the sludge being transported. It also discusses SC's refusal to accept waste. The cooldown of Unit II continues. (HP)

"Panel Urges N-Safety Steps." The ACRS calls for design and operation changes in nuclear reactors. These changes are expected to reduce a chance of nuclear breakdown. (HP and NYT)

"Crisis Aid Step Taken by Borough." Hummelstown Borough declares a state of emergency from 3/28 to the present. This places the town in line for any future federal or state aid granted due to the crisis. (HP)

"Poll Shows Readers Oppose N-Energy and TMI." TMI accident poll results tabulated by the MPJ are presented. In addition to the results and their explanations, the article also includes some remarks that were submitted with the poll responses. (MPJ)

"Radiation; We Know It's There But Where Is It Going?" The work of the Air Force's Sixth Weather Squadron at the Harrisburg International Airport is explained. The group's job was to track the radioactive plume emitted from TMI. (MPJ)

"Citizens Want to Act, Not Wait." The first regular monthly meeting after the TMI mishap of the Middletown Borough Council is reviewed. (MPJ)

4/18/79 "TMI Review Set."
A deputy fire chief of Middletown, Mel Hershey, gives views on the TMI mishap. Lack of teamwork was the fire chief's main complaint. Inadequate evacuation plans, insufficient personnel, and general confusion were cited as existing problems. (MPJ)

"Health Tests Once and Done?"
Public concern over long term radiation effects is examined. The reporter requests that the Federal Government and the NRC perform long range tests in order to better understand the effects of low level radiation. (MPJ)

"TMI Repairman Reminds Us No One Was Injured During Incident."
A review of the activities of one TMI repairman who believes that "no one was hurt or seriously overdosed" during the accident. (MPJ)

"No Radioactivity Reports From Ground, Surface Water."
No radioactivity was detected in any of the several water samples taken from points near TMI. The results indicated that the ground water was unpolluted, and people with wells had "nothing to worry about." (MPJ)

4/19/79 "Critics of Nuclear Power to March on Capitol on May 6."
Plans of a 5/6/79 antinuclear power coalition in D.C. are announced. (NYT)

"County Panel Opposes Extra Met Ed Costs."
Dauphin County Commissioners resolve to oppose passing Met Ed costs related to TMI on to the consumer. (HP)

Editorial: "Met Ed Solvency."
The editorial charges that Met Ed is going to great lengths to portray its financial problems. The editorial

states that the main issue is being neglected, i.e. the PUC is not charged with guaranteeing Met Ed's financial solvency, but must establish reasonable rates and adequate service for customers. (HP)

4/19/79 "Sen. Heinz Packs Then In; No Snap Decision on TMI."
Before an overflow crowd in WGAL TV's auditorium, Senator Heinz indicated that problems persist, but he is not ready to abandon nuclear power. (LIJ)

4/20/79 "PUC Freezes Met Ed Rates Depending Probe."
The PUC ruled that the cost of replacing power lost when the accident crippled the TMI plant will not be passed on automatically to consumers. It also froze a rate increase approved on 3/22 and ordered studies on the impact of the accident. (LIJ and HP)

"O'Pake Urges Protection for Utility Consumers."
State Sen. O'Pake introduced a bill to prevent utility companies from automatically passing to consumers the costs of nuclear mishaps. (LIJ)

"Accidents Fund for Utilities Urged."
NJ Gov. Brendan Byrne calls for creation of a compensation fund to cover costs of nuclear accidents such as occurred at TMI. (HP)

"Meetings Scheduled on N-Matters."
Seminars and local meetings are scheduled in the aftermath of TMI's accident. (HP)

"Iodine 131 in Air Drops."
The NRC reports that release of iodine gas has dropped as well as the concentration of the gas offsite. (HP)

4/21/79 "Reactor Coolant Cooler."
Unit II is brought closer to a shutdown as the temperature of coolant that circulates through the reactor drops 50 degrees. (HP and LIJ)

"TMI Area Resident Still Upset."
A Middletown resident, Alice Etter, was one of nine residents found to have higher than normal concentrations of radioactive material in her body. She plans legal action. (HP)

"N-Plant Training Queried."
Sen. Schweiker asserts that information from TMI indicates that the NRC's personnel efforts leave something to be desired. He calls for more stringent licensing and certifying of nuclear power plant personnel. (HP)

"Crisis Training Lined Up."
The NRC wants tougher training for plant operators so they can handle emergencies like TMI. Simulation training in the nuclear industry is discussed. (HP)

"13 Nations Pursue TMI Information."
Delegations of scientists, engineers, environmentalists, and politicians from 13 nations are in the US to study the nuclear accident at TMI. (LIJ)

4/23/79 "Florida Soil's Radiation Risk Exceeds TMI's."
Persons who live in homes on land rich in phosphate have a greater chance of dying from cancer due to radiation exposure than nuclear workers. (HP)

"Iodine 131 Emissions Fluctuating, NRC Reports."
Various Iodine 131 readings are reported and are described as within acceptable limits. (HP)

4/23/79 "Firm Unprepared, Says Met Ed."
Walter Creitz says the firm was not prepared for the 3/28 accident nor aware of its scope for two or three days. Creitz says that Met Ed proposed that the NRC serve as the only spokesman during the crisis. He contends that there was no pressure placed on Met Ed to do this. (HP, LIJ, and NYT)

"Doctor Offers Meltdown Prognosis."
Dr. Helen Caldicott, a pediatrician at Children's Hospital Medical Center in Boston, spoke to area groups. She projects what would happen medically if a meltdown occurs and talks of radiation related problems. (HP and LIJ)

"Denton Tapes TV Reply to Letter of Gratitude."
Denton taped a TV reply to a letter of gratitude from elementary students at Middletown School. (LIJ)

"Nuclear Tapes; Transcripts of NRC Emergency Meetings Concerning TMI Accident."
This report is based on taped emergency NRC meetings held during the incident at TMI. According to the article, the agency's commissioners and their assistants were themselves confused over the TMI events, partially due to the fact that "fast, reliable information from TMI" was hard to be had. Regarding the hydrogen bubble trapped in the reactor, the commissioners are reported to have been unsure of the danger. The article concludes with a brief description of Carter's eleven member committee organized to investigate the accident. Ann Trunk of Middletown is specifically noted. (NW)

4/24/79 "TMI Hearings Begin."
Hearings begin in Washington on TMI. Gov. Thornburgh is the first witness.

He sees the major problem as trying to get reliable information. Others assert that the public is being kept in the dark, while a utility spokesman believes that shareholders cannot be expected to pick up costs of the accident. (HP)

4/24/79 "TMI Builder Faulted."
The NRC considers a temporary shutdown of B&W plants. Problems in plant design are said to make them sensitive to malfunction and to place heavy demands on operators. (HP and LIJ; also 4/26/79 NYT)

"Senate Distracted by TMI."
Resolutions and bills related to TMI sparked activity in the Pa. Senate. Some call for politicians to stand aside and leave TMI concerns to technical experts, while others react to this. (HP)

"TMI Emission of Iodine 131 Reported Down."
Iodine 131 emission continues to drop. A new ventilation and filter system is to begin operation. This will cause a loud jet-like noise, but will help to lower levels of emission. (HP)

"Lawyer to Check on TMI Liability."
Cumberland County Commissioners hire a lawyer to research the Price-Anderson Act to determine if local governments can recoup funds spent on the TMI crisis. (HP)

"Met Ed Drops Talk for Pupils."
A Met Ed slide program scheduled for New Cumberland Middle School has been cancelled by Met Ed. No explanation is provided. (HP)

"TMI Waste Readied for Disposal."
A photo story of low level radioactive waste being transported to Hanford, Washington. (HP)

4/24/79 "Still Plenty of Bugs in Goldsboro Picnic Idea."
The idea of holding a post-TMI picnic in Goldsboro is debated pro and con. Some see it as saying the problem is over and are afraid that it detracts from the issue. (HP)

4/25/79 "Want to Voice Your TMI Concerns?"
The article includes a list of names and addresses of federal, state, and local officials to whom letters concerning TMI should be mailed. (MPJ)

"Steering Committee Hears Capitol Campus Programs Sparked by TMI."
The article reviews activities planned in response to the TMI accident at PSU, Capitol Campus. (MPJ)

"Medical Community Voices Concerns Over TMI."
Comments made by local members of the medical community in response to TMI are reported. (MPJ)

"Responses to TMI Public Opinion Poll Keep Rollin' In."
The article announces more results to MPJ's TMI Poll. (MPJ)

"Met Ed Invited to Discuss TMI."
TMIA invited Met Ed's Walter Creitz to appear in Middletown in order to discuss and debate TMI issues. The purpose of the meeting was to give area residents a clarification of the accident. (MPJ)

"TMI has Many Names in the Past."
The article discusses the different names that TMI has worn over the years. (MPJ)

"No Radioactivity Detected, Farmers Should Carry On."
The report states that there were "no detectable levels of radioactive materials in the soil or plant life" near

TMI. Farmers were advised to proceed with their usual cropping operations. (MPJ)

4/25/79 "No Radioactivity Found in Susquehanna's Fish."
Susquehanna River fish were not radioactively contaminated due to TMI. (MPJ)

"Hearings on TMI's Effects on Other Plants."
A report of a meeting held by a subcommittee of the NRC to discuss long term implications of the TMI mishap. (MPJ)

Editorial: "Senate Priorities."
The editorial urges the Senate not to leave issues of TMI to technical experts. It views legislature as one powerful forum that can address concerns of the public that might otherwise be ignored. (HP)

"Clean-Up Tab Juggled."
Gov. Thornburgh suggested that consumers and taxpayers may have to bear some financial burdens of the cost of TMI. He also made positive statements about the role of the press during the crisis. (HP and LIJ)

"Hearing Series on Met Ed Rates Slated."
The PUC schedules hearings on Met Ed price increases and controversy surrounding the public's liability for costs associated with the crisis. (HP and LIJ)

"More Instructions Due for Reactor Operators."
The NRC plans training for nuclear reactor operators similar to that given to nuclear submarine engineers. Reactor operators' jobs are discussed. (HP)

"Unit 2 Targeted for 'Safe Status.'"
The NRC plans for a cold shutdown of Unit II are detailed. (HP and LIJ)

4/25/79 "House OKs Panel to Probe State's TMI Response."
The Pa. State House created a panel to investigate the state's response to the nuclear accident at TMI. (HP)

4/26/79 Editorial: "Those Who Run Reactors."
The editorial alleges that the 'technocrats' are having too much say in whether TMI should be restarted. (LIJ)

"Experts Offer Varying Views on Radiation."
A seminar on low level exposure to radiation produced varying opinions on effects and possible consequences of radiation exposure associated with TMI. (HP)

"N-Closings Eyed."
The staff of the NRC recommends a temporary shutdown of eight power plants designed by B&W. (HP and LIJ)

"Agency Reports Turbine Mishap."
The NRC reports that a turbine had to be shutdown at Unit II. This resulted in modification of a system being used to cool the damaged reactor. The mishap and modification are explained. (HP)

"TMI Panel Set to Hear First Witness Thursday."
The Kemeny Commission on TMI is set to begin hearings. The first witnesses will be from UCS. Commission members, their pay, and benefits are listed. (HP)

"Suits Filed to Cover All TMI Damage."
Five class suits have been filed against Met Ed to cover damages for injury done to business property and for physical and mental harm. Persons within a 20 mile radius are covered in the suits. (LIJ)

4/27/79 "TMI Causes GPU to Cut Dividends."
GPU announced a reduction in its quarterly dividend from 45¢ to 25¢ a share in order to conserve cash reserves. (LIJ)

Editorial: "The Scene Changes."
For Pres. Carter to remain credible, the editorial suggests that he order the closing of all B&W reactors until they can be operated without any question of their safety. (LIJ)

"TMI Iodine Count Drops."
NRC officials reported that leaks of radioactive iodine are at their lowest levels in nearly two weeks. (LIJ)

"State Eligible for Nuke Crisis Grants."
Pa. is eligible for federal grants to study and solve economic problems created by the TMI nuclear accident. Grants are being issued by the EDA. (LIJ)

4/28/79 Editorial: "Good News and Bad."
B&W plants will be closed for safety checks. The editor also states that permanent closure of TMI may be needed to restore the integrity of the geographical area. (LIJ)

"NRC Accepts Compromise on Nuke Plant Shutdowns."
The NRC agreed to a temporary closing of nuclear reactors made by B&W, but let two such reactors operate for a time, a step that may prevent an electric power shortage in three southern states. (LIJ and NYT)

"Satellite may be Used to Detect Radiation in Water."
The USEPA announced that it might employ an earth-orbiting satellite as part of a warning system to detect any release of radiation into the Susquehanna River from TMI. (LIJ)

4/28/79 "TMI Tourist Fallout: Two Different Views."
Tourist officials (Lancaster) state that tourism business is rosy. However, a sampling of tourist-related businesses suggests that the accident continues to be a definite economic liability to some segments of the local hotel and tourist industries. (LIJ)

4/30/79 "TMI Plant Lures Streams of Sightseers."
Sightseers streamed in Sunday for a close-up of the TMI nuclear plant. (LIJ)

5/1/79 "Radioactive Water a New TMI Threat."
Spillage of radioactive water onto the floor of the reactor building has doubled since the accident. (LIJ)

"County a TMI Disaster Area."
Five counties surrounding TMI, including Lancaster, were designated economic disaster areas by the SBA. (LIJ)

5/2/79 "Improved Technology and Training in Nuclear Industry Urged."
Secretary Schlesinger says TMI underscores the need for improved technology and training of plant operators. He also stated that nuclear power remains essential for the American economy. (NYT)

"Township 'Cools Down' After TMI."
The article contains reviews of a Lower Swatara Township Board of Commissioners meeting. A few post-TMI plans were discussed, but the meeting was determined to be relaxed in the wake of the TMI accident. (MPJ)

"Time to 'Shovel-Out.'"
Unlike floods, fires, and blizzards, TMI cannot be cleaned up. The reporter asks the public to put TMI into perspective and to try to live with it for the sake of their mental and physical health. (MPJ)

5/2/79 "Students and Faculty Express Thoughts on TMI."
Students and faculty respond to the question, What are your personal feelings about the TMI incident? (MPJ)

"Met Ed: It Wasn't Too Bad."
Commissioner Myers attacked what he considered to be problems of management and arrogance on the part of Met Ed. He suggests Met Ed should evacuate the area. Pres. Creitz says the most important job continues to be meeting the electricity demands of area citizens. He states that not a single individual has suffered injury to his person or property as a result of developments at TMI. (HP)

"What Bankruptcy Could Mean for Met Ed."
In response to a PUC decision to freeze Met Ed rates at pre-TMI levels, the firm attorney predicted bankruptcy. (HP)

"Bubble at TMI was No Peril."
Roger Mattson of the NRC, who acted as an aide to Denton at the height of the TMI crisis, says he and colleagues misunderstood the peril of a hydrogen-oxygen explosion in the reactor. (HP)

Editorial: "More TMI Trouble."
The editorial alleges that the NRC is protecting Met Ed and nuclear energy rather than insuring public safety. (LIJ)

"In Bainbridge. Fears Still Linger."
In a town meeting, 300 Bainbridge residents asked pointed questions about the details of future evacuations. (LIJ)

5/3/79 "Air TMI Views, Public Urged."
Chauncey Kepford spoke at a teach-in on radiation at PSU, Capitol Campus. He urges the public to let the government know they oppose reopening Units I and II. (HP)

5/3/79 "New Suit Filed in Federal Court Over Reactor Accident."
A class action suit was filed in US Middle Court. The suit adds complaints and defendants to the suit filed earlier in Dauphin County Court. (HP)

"France Pushes N-Power Despite TMI."
An 11 member French delegation visited Harrisburg to discuss TMI with state and local officials. France will continue its development of nuclear energy despite the TMI accident. (HP)

"NRC was Aware of TMI Defect Before Accident."
Despite an acknowledged defect in the TMI units' emergency core cooling system, the NRC authorized continued operation of the plant prior to the accident. The defect apparently had no bearing on the accident. The defect and remedy are described. (HP)

"Thornburgh 'Jury' Debating N-Power."
Gov. Thornburgh addressed a meeting of the American Society of Newspaper Editors in NYC. The future of nuclear power in Pa. is unclear. Capsule remarks of other speakers are also cited. (HP)

"Help Sought by Met Ed on Revenue."
Met Ed asks the PUC to do something so that it can assure the financial community that it can repay loans. The company proposes a rate increase. (HP)

"SBA Sets Up TMI Shop Here."
An office was set up in the Lancaster County Courthouse to handle requests for TMI related business loans. (LIJ)

5/4/79 "Califano Revises Nuclear-Toll View."
HEW Secretary Califano says radiation exposure from the TMI accident reactor was higher than earlier measurements indicated; high enough to cause 1 to 50

cancer deaths among 2 million people living within 50 miles of the plant. Califano's remarks were testimony before the Senate Subcommittee on Energy. (HP, LIJ, and NYT)

5/4/79 "What to Do With Water Keeps TMI in Quandry."
A way to dispose of the radioactive water in the containment building plagues NRC officials. (LIJ)

"Penn State Educator Briefs TMI Probe Panel."
Dr. Warren F. Witzig explained operations, hazards, and other aspects of nuclear reactors to a Pa. House Select Committee on TMI. (HP)

"City Sends Carter a Message on TMI."
The Harrisburg City Council resolution calling for TMI to remain inoperative until fail-safe methods of operation have been developed and sent to Pres. Carter. (HP)

5/5/79 "Tag Theory Linked to Accident at TMI."
A Met Ed official believes that a tag obscuring two lights on the Unit II control room panel may have compounded the accident. (HP)

"Panel to Oppose TMI Reopening."
PANE is formed at a public meeting of Middletown area citizens. Formation is announced by John Garver, Jr. (HP)

"Governor Approves Plan to Dump TMI Water."
Gov. Thornburgh has approved a plan where decontaminated radioactive water can be dumped into the Susquehanna River. (LIJ)

"Concern, Not Panic: No Rush for Abortions After Nuclear Scare."
Private and state medical authorities appear to have allayed the fears of

pregnant women here. There is no rush for abortions. (LIJ)

5/5/79 "Carter Politics in Iowa, Assails Foes." During an Iowa news conference, Pres. Carter says he hopes TMI will not lead to the abandonment of current nuclear technology and to the development of breeder reactors. (NYT)

"Moves Across Country of Radioactive Sludge Stir Ire in Some States." Low level radioactive sludge from TMI is being shipped to Hanford, Wash. by truck. Local officials along the truck route were not notified in advance by the NRC. (NYT)

5/6/79 "A Year's Ban on Access to Plant." A GPU spokesperson states 2 to 3 years will be needed to decontaminate the plant. (NYT)

"Siting Nuclear Reactors Once Seemed Simple and Safe." The article expresses concern over the proximity of nuclear power plants to population centers in light of the accident at TMI. (NYT)

5/7/79 "65,000 Demonstrate at Capitol to Halt Atomic Power Units." At least 65,000 people march on the Capitol on 5/6 in the largest demonstration against nuclear power the US has ever seen. (NYT and HP)

"Herbein Offers Accident Cuase." John Herbein told a Pa. House Committee that the accident at TMI was not detected right away because there are no signals on the control panel to indicate that a pressurized relief valve is stuck open, and is allowing the coolant to escape. (HP)

5/7/79 "Met Ed to Cut Costs for Most May Users."
Most of Met Ed's customers will pay less for electricity due to adjustments in the cost adjustment charge and the state tax surcharge. (HP)

"A Nuclear Shutdown."
The problems encountered by the NRC while struggling to decide whether or not to shutdown all seven nuclear reactors that are similar to TMI are discussed. To shutdown the plants for safety's sake could mean a serious power shortage for millions of people. The strong views of some political leaders are presented. (NW)

"Fallout for B&W; TMI Reactor Built by B&W."
The financial and legal woes of the B&W Co., the builder of TMI, are discussed. Liabilities from the TMI accident are predicted to be a major problem to the company's financial situation, and these liabilities for personal property damage may not surface for months or even years. Also, the negative publicity from the TMI incident and other alleged action appears to be seriously damaging the company's image, and consequently sales of B&W's other conventional power plants are expected to dwindle. The company's stock is reported to have dropped 23%. (NW)

Editorial: "What if Nobody Calls."
The editorial opposes the dumping of decontaminated water into the Susquehanna River. (LIJ)

"City to Shut Filter Plants if TMI Water Discharged."
Lancaster Utilities Superintendent, Mike Freedman, stated that the city will temporarily close its water plant on the Susquehanna River if advance notice is given when decontaminated water from the

containment building at TMI is discharged in to the river. (LIJ)

5/8/79 "Walker Writes NRC, Protests Dumping Water."
Cong. Walker has written to NRC Commissioner Hendrie protesting plans to dump radioactive water from the crippled TMI plant into the Susquehanna River. (LIJ)

"Cleanup of TMI Water Starts Soon."
The NRC stated that efforts to cleanup TMI radioactive water could start by the end of May. (LIJ and HP)

"Met Ed, NRC Knew Core Damaged 2 Days Before it was Reported."
Control Room Supervisor, Jim Floyd, told visiting congressmen that control room personnel and federal inspectors had the information that the plant's fuel core was seriously damaged two days before it was formally reported. The congressmen were members of the House Committee on Energy. (HP, LIJ, and NYT)

"Carter Tells Protesters Closing All Nuclear Plants is 'Out of the Question.'"
Pres. Carter says minimizing the need for nuclear power is policy, not an abandonment of nuclear energy. (NYT)

"Crew Praised for Evac. Readiness."
A federal official praises Cumberland County for service during the period when an evacuation was a possibility. (MPJ)

"Pennsylvanians Lobby for N-Shutdown."
The article contains the comments on Pa. participants in a D.C. antinuclear demonstration. (MPJ)

5/9/79 "TMIA Announces Studies to Pursue."
TMIA is compiling information in order to determine the immediate and long range effects of the TMI crisis.

Individuals who had premonitions and para-normal experiences prior to the accident will be studied. (MPJ)

5/9/79 "NRC Advisory Committee to Meet in Washington."
The schedule of a meeting to be held 5/10-12 in D.C. to discuss the TMI mishap, by the NRC Advisory Committee is listed. (MPJ)

"PUC Gets Earful From Let Met Ed Pay Advocates."
The public voices opposition to Met Ed passing on costs of accidents to consumers. (HP and LIJ)

"Met Ed Offers Plan to Recover Costs."
Met Ed asks the PUC for two additional temporary increases to cover TMI costs 10-16 months prior to the accident. (HP and LIJ)

"Fate's Quirk: TMI Disaster Potential Described by Radiation Scholar."
John W. Goffman, antinuclear physicist, said that if the TMI nuclear accident had a larger inventory of nuclear fuel, the accident may have been worse. (HP and LIJ)

"Plans to Dump TMI's Water Upset Council."
Lancaster City Council members expressed grave concern over the plan to deposit large amounts of radioactive water into the Susquehanna River. (LIJ)

"Water Dump Plan Opposed by Wohlsen."
Mayor Wohlsen, in a letter to Gov. Thornburgh, indicated his opposition to the plan to dump decontaminated water into the river. (LIJ)

"Deadline for Nuclear Waste Urged."
Senator Hart, in a speech before the National Press Club, says that if an acceptable plan for nuclear waste is not

adopted in five years, all nuclear plants should be shutdown. (LIJ)

5/10/79 "House Panel Votes Limits on A. Plants." House Interior Committee approves a six month moritorium on construction permits for nuclear reactors and a ban on the operation of new reactors until the federal government develops a plan to deal with nuclear accidents. (NYT)

"Unit 2 Core May Be Lost."
A GPU official tells the PUC that the core of the Unit II reactor may have to be scrapped. A similar reactor could cost $88 million. (HP)

"TMI Incident Hearings Due at Middletown."
The Kemeny Commission will visit Middletown for a three day hearing. The public is invited to testify for five minutes per person. (HP)

"GPU Cuts Work Force 5% in Effort to Stay Solvent."
GPU, a part owner of TMI, cuts its work force 5% to stave off insolvency. A factor in their financial problem is the cost of paying for replacement power; $800,000 per day. (LIJ)

5/11/79 "PUC Denies Met Ed Surcharge Request."
A Met Ed surcharge request for '77 and '78 expenses to the amount of $16.3 million was denied by the PUC. (LIJ and HP)

"SVA to Ask Court to Block Water Dumping."
One hundred fifty members of SVA voted to legally block the dumping of decontaminated water. (LIJ)

"Indicator Slip Eyed in TMI Context."
Prior to the accident, B&W admitted that under some circumstances a control room

instrument could mislead operators to think that the reactor's core was covered with cooling water when it was actually exposed and in danger of melting. (HP)

5/11/79 "PP&L N-Plant Review Planned." Following the TMI crisis, PP&L has formed committees to review the Susquehanna Steam Electric Station under construction near Berwick. (HP)

"TMI Startup 'Backlash' Seen." Clifford Jones, Secretary of the Pa. DER, warns that a restart of Unit I could provoke the most massive demonstration in some time. (HP)

"TMI Crisis Evacuation View Aired." Oran K. Henderson, Director of PEMA, tells a Pa. House Committee that he recommended a general evac. to Gov. Thornburgh based on the emotional nature of a telephone call from a frightened TMI supervisor. (HP)

"Panel Oks TMI Probe Support Grant." The US House Public Works Committee approved a $400,000 two year grant to the EDA to study the economic impact of TMI on a five county area. (HP)

"Panel Votes Restrictions on Plants." Senate Environment and Public Works Committee votes to revoke the licenses of all nuclear reactors operating in states that do not develop federally approved emergency evacuation plans within six months. (NYT)

5/12/79 "TMI Could Have Been Just a Minor Incident." The NRC staff reported that a minor incident was turned into a major one because operators at TMI were unable to tell what was happening inside the nuclear reactor. The report also listed six human errors which compounded the problem. (HP, LIJ, and NYT)

5/12/79 "TMI Becomes Attraction for Tourists."
Pa. Secretary of Commerce, Bodine, believes that with the right promotion TMI can become an attraction for tourists. (LIJ)

"City Jumping in With a Lawsuit Over TMI Water."
Mayor Wohlsen announced a law suit on behalf of the City of Lancaster over the dumping of water from TMI. Wohlsen believes that no water should be discharged until the safety issues are addressed. (LIJ)

"TMI Water Discharge is Opposed."
A Pa. House Committee urges Gov. Thornburgh to work to prevent the discharge of water from TMI into the Susquehanna River. (HP)

"State Police Deny 'Probe' of TMI Protesters."
Bruce Smith, Chairman of Newberry Township Supervisors, reacts to what he believes to be a Pa. State Police investigation of the township's TMI Steering Committee. The state police commissioner says he knows of no such investigation. (HP)

5/15/79 "Ohio Route as N-Route is Denied."
An Ohio Turnpike official denies reports that radioactive wastes from TMI were secretly transported across the Ohio Turnpike. (HP)

"NRC Installing Hotline Linkup."
The NRC acts in the wake of TMI to install telephone hotlines to link the nation's 70 commercial nuclear power plants with agency headquarters. The NRC declines to recommend that the federal government should take over plant operation following accidents such as TMI. (HP)

5/15/79 "GPU Gets State Tax Refund."
GPU received a $13.9 million state tax refund covering 1971-75. The refund will be divided between Met Ed and the PEC. (HP)

"NRC Says Realization of Blast was Unlikely."
NRC inspectors denied a Met Ed supervisor's claim that the NRC should have known in the early hours of the crisis, that an explosion had occurred within the Unit II reactor containment building. (HP)

"Gov., NRC Discuss Dumping into River."
Rumors continue that the NRC will immediately dump water into the Susquehanna River. The NRC told the Governor that the highly contaminated water at TMI will not be discharged into the Susquehanna River in the near future. (LIJ)

"Anti-nuclear Groups to March on Met Ed."
Met Ed headquarters will be the scene for an antinuclear protest march on 5/20. (LIJ; also 5/17/79 HP)

5/16/79 "Contaminated Work Clothes Found in TMI Trash Dump."
Between 15 and 20 plastic bags holding contaminated work clothes were found buried in a trash dump at TMI. Both the NRC and Met Ed officials said that there was not a health hazard despite the improper disposal of the clothes. (LIJ)

"Health Agency to Hear Woes by TMI."
A schedule of hearings concerning TMI were announced by the regional health systems agency. (LIJ)

Editorial: "The Ultimate Test."
The editorial criticizes the limitations of liability of the Price-Anderson Act and urges voters to write Cong. Walker to take action to extend liability limitations. (LIJ)

5/16/79 "Stockholder's Suit Alleges TMI Deceit."
A GPU stockholder files suit alleging that the company deceived stockholders by failing to warn them of the potential for a TMI accident. (HP and NYT)

"TMI Recovery Topic of Session."
Cumberland County Commissioners will hold a conference to launch a local recovery project to counter the effects of the TMI accident. (HP)

"Legacy Seen in TMI: Public Demands to be Fully Informed."
Joyce Treeman, Executive Director of the Pa. Commission on TMI, says the accident altered public perception of the government's role as a regulator. (HP)

"Loans to $100 M Available."
An office was opened in Middletown for the purpose of providing long term loans to area small businesses who claim economic injury as a result of the TMI accident. Details and instructions concerning the loans are given. (HP)

"TMI Lecture Series Scheduled at HACC."
HACC presented a series of lectures dealing with the subject of TMI. Schedules for these lectures are given. The reported purpose for the lectures was to present a balanced view of the accident and its past and potential effects on the area. (HP)

"Commissioners Prepare Survey to Take Stand on TMI N-Issue."
The article includes a report of a Lower Swatara Township Board of Commissioners meeting. The attending members decided that citizen input in the form of reactions, requests, suggestions, and general feelings would be instrumental in determining the fate of the disabled power plant. Plans for the preparation of a questionnaire survey were made. (HP)

5/16/79 "Med Center's Workshops Aided Many to Understand and Cope with TMI."
A series of workshops were sponsored by the HMC for its employees for the purpose of helping them deal with the effects of stress from the TMI crisis. The HMC hoped that these programs would help guide employees toward a renewed state of normalcy. (HP)

5/17/79 "TMI Probers will Convene."
The Kemeny Commission will convene with Gov. Thornburgh as the lead witness. (HP)

"TMI Area Due Health Profile."
PHD will use a $300,000 grant to begin health profiles of 50,000 persons within a five mile radius of TMI. (HP)

"TMI Probe 'Like Detective Story.'"
Presidential Commission Chairman, Kemeny, is profiled. (HP)

"Unit 2 Water Dumping Held Up."
The NRC promises Pa. and local officials that radioactive water from Unit II will not be dumped into the Susquehanna River unless it is cleaned up and meets federal standards. (HP)

"City Will File Suit Despite Assurances TMI Water is Safe."
Top Met Ed and NRC officials met with city officials to convince them that the drinking water of Lancaster will not be harmed by the dumping plan. City officials were unconvinced and they plan to continue with the legal suit. (LIJ)

5/18/79 "Idea for Monitoring TMI by Eye-In-Sky is Abandoned."
Because the proposed method provides a 15 minute sampling rather than continous monitoring of TMI, Pa.'s DER and USEPA have decided to abandon the plan for satellite monitoring. (LIJ)

5/18/79 "Nuke Accident Loan Office Cutting Hours." The SBA will reduce its special disaster hours from six days a week to two days because the volume of business has been decreasing. (LIJ)

"TMI Probers Call Off Hearings." The Kemeny Commission votes against hearing testimony from public officials or utility representatives until Congress grants them subpoena powers and the authority to swear in witnesses. (HP)

"TMI Unit Subpoena Bill Gains." The Senate passed a bill granting subpoena power and immunity for witnesses to the Kemeny Commission. (HP)

"N-Insurance Called U.S. Folly." Cong. Goodling told a public meeting in Newberry Township that if the federal government removed itself from insuring nuclear reactors, suppliers would be forced to develop alternate, safer power sources. (HP)

"2 TMI-Related Bills Illegal, PUC Aide Says." A PUC official called two bills in the Pa. House unlawful. Both bills deal with rate increases and how to enact them. (HP)

"Cooling-Water Testing Faulted in N-Accident." Government officials believe that the thrice a month cutting off of the TMI reactor's supply of cooling water, as a test procedure, contributed to the plant accident. (HP)

"GAO Says NRC Bungled Warning About TMI." A regional inspector believed in January that there were safety problems with B&W designed plants. The ASLB was not informed of the inspector's concern until the day after the accident. (HP; also 5/19/79 LIJ, and NYT)

5/19/79 "State Monitoring at TMI Behind its Schedule."
DER monitoring of the place where waste water would be dumped into the river has not yet been put into operation and is behind schedule. This system is designed as a back-up to Met Ed's own monitors. (HP)

"19 Cows Die Mysteriously on Farm 4.5 Miles from TMI."
Since the accident, 19 cows have died mysteriously on a local farm. A report of the Bureau of Animal Industry Laboratory in Summerdale indicated that there is no evidence that the deaths are related to TMI. (HP)

"Panel Probing TMI to Hear From Public."
The Kemeny Commission will take 10 minute testimonies from the public. A bill is pending giving the panel authority to subpoena witnesses and place them under oath. (HP; also 5/20/79 NYT)

"NRC Defends Educational Requirements."
The NRC challenges Sen. Schweiker's contention that high school graduates should not serve as control room operators in nuclear power plants. (HP)

5/21/79 "Anticipate Trouble: Denton."
Harold Denton tells the LVC graduating class that TMI taught us trouble must be anticipated and plans should be made to cope. (HP and LIJ)

"Protestors Gather at Met Ed's Front Door."
A protest at Met Ed's headquarters in Reading attracted over 1,000 persons. (HP, LIJ, and NYT)

5/22/79 "Unit to Monitor Studies on TMI."
A volunteer group formed as an ad hoc steering committee for Cumberland County Commissioners will monitor studies of agencies looking at the impact of TMI. (HP)

5/22/79 "Neighbors Vow to Close N-Plant."
Residents packed Middletown Borough Council's meeting to demand action to permanently close TMI. (HP)

"Air Over TMI Finally Clear."
Tests indicate the air over TMI is finally free of radioactive traces. (HP)

"Cattle Death Radiation Discounted."
Radiation is discounted as the cause of death of 19 cows on a farm close to TMI. (HP; also 5/23/79 LIJ)

"NRC Asked to Bar Dumping of 'Hot' Water."
The Harrisburg CSE and the York CSE have petitioned the NRC to prevent the dumping of processed radioactive waste water from Unit II into the Susquehanna River. (HP)

"NRC Suit Filed by Lancaster."
The City of Lancaster asks a federal court to bar the discharge of radioactive waste water from TMI into the Susquehanna River. (HP and LIJ)

"Faulty Equipment Blamed by Panel."
Cong. James Weaver of Oregon asserted that equipment and instrument failure played a greater role in the TMI accident than operator error. He asserted that the same type of accident could occur at other N plants in the future. (HP and NYT)

"NRC to Hold Off on New N-Licenses."
The NRC will delay the issuing of licenses for new N plants for at least three months until lessons learned from TMI can be implemented. (HP and NYT)

"TMI Partner to Use Uranium as Collateral."
JCPL Co., a partner in the operation of TMI, won permission of the NJ PUC to use stockpiled uranium as collateral for loans to head off threatened bankruptcy. (LIJ)

5/23/79 "Citizens Demand Councils' Views on TMI." The article contains reviews of a meeting held between PANE's John Garver and the Middletown Borough Council. (MPJ)

"Commission Gets Close-Up of TMI." This report describes a few of the activities of the Kemeny Commission including a tour of Unit I. (MPJ)

"Graduation Will Proceed Despite TMI Crisis."
The article outlines the provisions of a bill that amends the Public School Code so that seniors could graduate as scheduled despite class days lost due to the TMI mishap. (MPJ)

"Myers Will Testify on N-Mishap Plans." Commissioner Jacob Myers and others will testify before the US House Armed Forces Committee which is examining disaster response capabilities in connection with the TMI accident. (HP)

"Met Ed Credit Crunch Blamed on PUC." New York banks assert that a lack of confidence in what the PUC will recommend for a Met Ed rate hike is a block to lending Met Ed funds. (HP)

"Speakers Say Met Ed Should Pay the Price."
Speakers in a TMI lecture series at HACC say the nuclear industry and Met Ed should carry the economic burden of TMI. (HP)

"Utility's Fate Debated."
NJ state officials debate the future of JCPL and its request for a rate hike. (HP

"TMI Panel Named."
Gov. Thornburgh names a panel to study economic, environmental, and health effects of TMI. Lt. Gov. Scranton will head the committee. (HP)

5/23/79 "Radiation, Cow Unlinked."
The Pa. DOA finds no evidence that radiation caused the death of a Lancaster County cow. (HP; also 5/24/79 LIJ)

"Panel Probes Health Aspects of TMI Accident."
A panel from HRPD conducted an informal hearing of TMI area residents. Citizen comments are provided. (HP; also 5/24/79 LIJ)

5/24/79 "CAT Puts TMI Loss at $11,500."
CAT sustained $11,500 in losses as a result of the TMI crisis. (HP)

"GPU Ready to Make Final Dividend Decision."
The GPU Board of Directors will meet to decide if it will pay a common stock dividend. The money cannot be paid legally unless the corporation is solvent. (HP and LIJ)

5/25/79 "Profs Give Thanks to Met Ed."
A University of Pa. professor told a York farm group that Met Ed should be complimented for containing radiation at TMI. Others state that there was no damage to farm animals, milk, or crops in York County as a result of TMI. (HP)

"No Cover-Up, Met Ed Aide Says."
Herbein says there is no evidence that the company tried to keep the NRC in the dark about the TMI crisis. (HP)

"Met Ed Customers Mail in Protests."
The PUC received more than 4,000 letters protesting any plan to have consumers pay for the TMI nuclear accident. The mailings were organized by the Utilities Consumer Group, formed in April by Ed Boas, a Bucks County farmer. (LIJ and HP)

5/25/79 "TMI Suit to be Filed Today."
Lawyers for the SVA intend to file suit in a Harrisburg federal court to stop radioactive water from being released into the Susquehanna River. (LIJ and HP)

"Cattle Herd AOK at Marietta Farm."
A nearby farmer reports his cattle are OK and he doesn't understand the concern over the death of 19 cows at another farm close to TMI. (LIJ)

5/26/79 "City Wins 1 Battle with TMI."
The NRC banned the discharge of radioactive water from TMI into the Susquehanna River until its staff prepares an environmental statement. SVA suit for preliminary injunction denied due to above voluntary action of NRC. (HP, LIJ, and NYT)

"NRC Rep to be Asked to Attend TMI Pre-Trial Sessions."
The NRC will be asked to send representatives to a pre-trial conference on TMI suits. Class action suits are explained. (HP)

5/29/79 "Met Ed Plea Asks End to Rate Hike Freeze."
Met Ed petitioned Commonwealth Court to set aside a PUC order on a previously approved rate hike. (HP)

"President of Dartmouth Turns Into N-Sleuth."
The article gives a profile on Dr. John Kemeny, Chairman of the Presidential Commission on TMI. (HP)

5/30/79 "Subpoena Power Bid Fails."
A bid by the Pa. House Committee investigating TMI to gain subpoena power failed. (HP)

5/30/79 "Met Ed Hike Could Be a Shock."
The possible costs to Met Ed customers paying for power lost at TMI is discussed. (HP)

"Court Order Issued in Water Suit."
The US District Court in D.C. issued an order binding the NRC to its temporary ban on the discharge of radioactive water from TMI. (LIJ)

"HEW to Study Accident Effects in A-Plant Area."
HEW and PHD will undertake extensive studies of the health effects of the accident at TMI for people living within five miles of the plant. (NYT; also 5/31/79 MPJ)

5/31/79 "A-Plant Aides Describe Previous Errors at Key Valves."
M.V. Cooper and E.D. Hemmila, control room operators at TMI, testify before the Kemeny Commission, that valves were mistakenly closed at the plant on a number of occasions before such an error crippled the plant on 3/28. Kemeny and Herbein disagree over the degree of sophistication of computer technology at TMI. (NYT, LIJ, and HP)

"Bill Puts N-Accident Cost on Utility."
Pa. Senate passes a bill that consumers will not pay for N plant accidents if the utility company is at fault. (HP, LIJ, and MPJ)

"Groups 'Disappointed' by Governor on TMI."
Antinuclear groups note dissatisfaction with the Governor following discussions on TMI. (HP)

"Two Would Let Met Ed Pay."
Government witnesses recommend to the PUC that costs associated with TMI be excluded from the new revenue stucture for Met Ed. (HP)

5/31/79 "Boro Takes Long, Hard Look."
A committee met to formulate an Emergency Preparedness Plan for the TMI area. (MPJ)

"TMI Resolution Falls Short in Township."
The article reports on the proceedings of a Lower Swatara Township meeting on TMI. A resolution was proposed to monitor area residents' health to determine long range effects of radiation. The resolution was voted down, and each Commissioner was asked to alter the resolution as he saw fit for discussion at a later meeting. (MPJ)

"Met Ed Offers Workers Voluntary Retirement."
The article announces the details of a voluntary early retirement plan for Met Ed employees. The plan was part of efforts to shore up their financial position in the wake of the accident at TMI. (MPJ; also 6/1/79 HP)

"Subcommittee Meetings on TMI Here."
A subcommittee of an NRC committee was scheduled to conduct a meeting in Middletown to discuss technical details of the TMI mishap. (MPJ)

"Showed Significant Percentages Fear Results of TMI Incident."
The results of a Michigan State University survey are reported. Residents within a 20 mile radius of TMI were studied three weeks after the mishap. The results show that these residents feared the disaster and expected an impact on their lives. (MPJ)

"Who Are the Commissioners Studying TMI?"
A brief biography including the credentials of the members of the Kemeny Commission is presented. (MPJ)

6/1/79 "TMI-I August Startings Eyed."
Met Ed aims to startup Unit I by 8/79. (HP)

"GPU Reports Net Income Drop of 4% Over a Year."
GPU reports a 5% loss of income for the first four months of 1979, and 4% over a twelve month period. The amount of loss due to TMI is said to be undeterminable at the present time. (HP)

"NRC Scrapped Evacuation."
Denton had recommended an evacuation on 3/30, but rescinded it within an hour, feeling there was an initial overreaction. He also stated that the NRC staff was somewhat complacent about reactor safety before the accident. (HP and NYT)

"No Make Up Days Because of TMI."
School districts which lost days because of the TMI nuclear accident will not have to make up those days, according to a new law signed by Gov. Thornburgh. (LIJ)

6/2/79 "GPU Backs Plan to Up Rate 11.2%."
Pa.'s Office of Consumer Advocate advanced a plan to allow Met Ed to recoup $41.5 million from customers over an 18 month period as a return payment for fuel purchases. Met Ed's customers would pay 11.2% more than before TMI. (HP)

"TMI Study Stirs Ire of Probers."
Some members of the ACRS express criticism of B&W actions and lack of actions to prevent future accidents like TMI. (HP)

"Hershey Electric Fights Rate Hike."
The HEC asks the FERC to prevent Met Ed's proposed rate hike for wholesale

customers from taking effect. Hershey contends that the TMI crisis is the main reason for the hike which will increase residential customer's bills by $9 a month. (HP)

6/2/79 "TMI Exempt From Rules."
TMI had been exempted from some federal rules on plant design because it was under construction at the time of adoption. The nature of the rules and possible application to TMI are discussed. (HP and NYT)

6/4/79 "Governor Calls Denton a Hero."
Thornburgh says the real hero of TMI was Harold Denton. He urges graduates at Bucknell University to follow Denton's example and to do well in whatever they are trained to do. (HP)

6/5/79 "Still Unsure About Role, N-Regulators Admit."
NRC members state they are confused about their roles in nuclear emergencies. As a result of TMI, staff procedures in responding to accidents have been strengthened, but roles have not been sorted out. (HP)

Editorial: "Unit 1."
The editorial expresses the opinion that until investigations relative to Unit II are completed, Unit I should not reopen. (HP)

"Pioneer Urges N-Power Consortiums."
The nuclear power industry must be consolidated into a handful of large consortiums running a few, huge nuclear plants, if it is to survive the fallout from the TMI accident. The speaker was Dr. Alvin Weinberg, Director of the Institute for Energy Analysis, Oak Ridge Associated Universities, and formerly Director of the Oak Ridge National Laboratory for 25 years. (LIJ)

6/6/79 "B and W Says Met Ed Erred."
B&W denies any blame in the TMI accident. The company believes that mistakes made by Met Ed control room operators, not equipment or design, led to the crisis. (HP, LIJ, and NYT)

"Another Panel Sets Up Shop to Probe TMI." Gov. Thornburgh sets up a Commission to study and evaluate the consequences of the incident at TMI. The panel will study economic, environmental, and health effects only. Pa. Health Secretary, Gordon MacLeod, also names his own panel to study health effects. (HP)

6/7/79 "N-Plant Halt Declined."
The federal House Commerce Committee fails to follow other panels in recommending a moratorium on N plants due to TMI. It also endorsed a proposal to require the NRC to notify states when it plans to truck radioactive wastes from TMI or other nuclear plants across their borders. (HP)

"GPU Attacks Theory on Radioactive Waste."
A GPU official disputes a previous theory of how water in Unit II auxiliary building became contaminated. The actual process is still labeled a mystery. (HP)

"TMI Crisis 'Bungled,' Residents Tell Panel."
Goldsboro residents give views on the TMI crisis to a Pa. House Select Committee. Confusion and conflicting information are cited. (HP)

"PP&L Will Sell Power to Met Ed."
PP&L has asked the PUC to approve its plan to sell replacement power to Met Ed due to the TMI accident. (LIJ)

6/7/79 "Goodling Asks Tax Credits for TMI Surcharges."
Cong. Goodling introduced a bill in Congress that would give federal tax credits to people whose utility costs have increased because of the TMI nuclear accident. (LIJ)

"GPU Struggles for Financing."
The accident at TMI has placed GPU in a financial squeeze between utility regulations in Pa. and NJ and banks. (NYT)

6/8/79 "Electric Rate Hike Delay Asked."
HEC asks the PUC for permission to postpone Met Ed's rate increase for wholesale customers. Hershey feels the increase to wholesale customers involves Unit II. (HP)

"'Worst to Come' from TMI Crisis, Activist Believes."
Chauncey Kepford tells a college group that the greatest radioactive dangers will come when the Unit II reactor building is physically entered and cleanup begins. (HP)

"TMI Testimony About Radiation Backs Met Ed."
Pa. radiation experts believe that Met Ed released factual information on radiation levels at Unit II. This testimony was given before a House Select Committee on TMI. (HP)

"Unit 1 Rated Ripe for Trouble on March 27."
A Met Ed official says that the valve mistakenly left closed on Unit I on 3/27 could have triggered an accident similar to that in Unit II. The article contains statements made before the NRC. (HP)

6/9/79 "Nuclear Industry Begins Study."
The N-power industry begins a $4.8 million effort to study the accident at TMI and convince the public that atomic power is safe and necessary. (NYT)

"NRC Blamed by Mayor Reid."
Mayor Reid blames the NRC for the TMI affair. He contends the plant was licensed before it was ready and that there were no inspectors at a plant with the potential to kill thousands. The testimony was given before a group of state legislators. (HP)

6/12/79 "Met Ed Bill Still Disputed, Method OK'd."
Agreements were reached on a method of imposing Unit II crisis costs, but groups differ on the dollar amount. (HP)

"NRC Weighing Start-up of Undamaged TMI Unit."
The NRC met with Met Ed officials to consider starting up the undamaged reactor. A GPU spokesman indicated that each reactor at TMI costs about $400,000 a day when it is out of service. (LIJ)

Editorial: "Too Soon to Consider Starting Up Unit One."
The editorial criticizes attempts by Met Ed to startup Unit I at TMI. The editorial is critical of Met Ed's failure to appologize for the accident. (LIJ)

6/13/79 "Census Set to Gauge TMI Effects."
The first phase of PHD's census of people living within a five mile radius of TMI will begin. Administration of the survey is discussed. (HP)

"Commissioners Tour TMI, Hear 'Clean' Water Plans."
Dauphin County Commissioners are told of plans to clean radioactive water at the TMI plant and eventually release it into the Susquehanna River. (HP)

6/13/79 "Panel Shapes Health Studies of TMI Crisis."
Several long term studies to assess the health effects of TMI are being planned by the PHD. (HP)

"'Evac.' Data Termed 'Confusing.'"
James H. Fisher, Executive Director of Emergency Health Services of South Central Pa. contends 'mass confusion' among emergency officials occurred in the wake of TMI. He criticizes the flow of information. (HP)

Editorial: "Reactor Unit 1."
The editorial expresses the belief that GPU is acting too matter-of-fact in trying to reopen Unit I. The editor feels the company discounts local fears. (HP)

"Health Official Claims News Accounts of TMI were Incorrect; Asserts There was No Real Danger During the Accident."
The Director of the BRP, Thomas Gerusky, asserted that the news accounts of the TMI mishap were incorrect. He stated that there was never a danger or possibility of hydrogen explosion, and that the evacuation of pregnant women and preschool children was unnecessary. Mr. Gerusky had additional comments concerning the environmental impact of the accident. (MPJ)

"City Waits for Ruling on Petition."
A federal ruling may be issued on Lancaster's petition to prevent discharge of radioactive water into the Susquehanna River from TMI. (LIJ)

"Walker Says TMI Start Premature."
Cong. Walker said that NRC's informal discussion with GPU concerning the start-up of TMI may be premature. (LIJ)

6/13/79 "38 Volunteers to Clean Up Plant at TMI." Thirty eight volunteers from PP&L and PEC respond to a call from Met Ed for assistance in the cleanup. (LIJ)

"Columbia's Mayor, Council Visit TMI." The mayor of Columbia stated that his visit to TMI did not alter his opposition to the discharge of radioactive water into the Susquehanna River. (LIJ)

6/14/79 "Death of Cows Near TMI Ruled Virus, Not Radiation."
State agricultural officials found a virus killed 19 cows at a farm in Bainbridge. (LIJ and HP)

"L. Swatara Resolution Demands TMI-1 Stay Shut Down."
Lower Swatara Township passes a resolution that says until the Kemeny Commission evaluates the results of the accident at Unit II, Unit I should remain closed. (HP; also 6/20/79 MPJ)

6/15/79 "Another Probe of TMI Set."
The NRC hires an independent Washington law firm to investigate its role at TMI and the accident itself. The firm is headed by Mitchell Rogovin. The investigation will not duplicate others. (HP; also 6/17/79 NYT)

"Methodists Seek N-Plants Curb."
The Central Pa. Conference of the United Methodist Church called for a moratorium on the construction of all nuclear power plants. It follows a similar call by the Central Synod of the Lutheran Church in America. (HP and LIJ)

6/16/79 "PUC OKs Hikes for Met Ed Penelec."
The PUC authorizes passing on supplemental fuel costs to Met Ed customers. This also applies to Penelec which owns 25% of the damaged unit. Met Ed owns 50%.

The utilities contend the effect of the hike is minimal and the rates are the same as those that would be in effect if TMI had not happened. (HP and LIJ)

6/16/79 "Red Cross Reevaluating Evac. Plans." Red Cross Pres., George M. Elsey, indicated that the Red Cross is reviewing its disaster plans in light of TMI as it had taken lightly the possibility of such an accident. (LIJ)

6/19/79 "TMI Vows No Startup without Public Okay." Herman Dieckamp stated that the owners of TMI nuclear plant won't startup the undamaged reactor at the plant until neighboring communities are convinced that it can be operated safely. (LIJ and HP)

6/20/79 "43 Banks Grant GPU $409 Million Credit." Wm. G. Kuhns stated that this credit should aid the utility through 1980, while its financial standing is being restored. (LIJ and HP)

"Suit Seeks TMI Billing Protection." Cong. Eugene Atkinson files suit in US District Court in Pittsburgh to prevent GPU from passing the costs of the nuclear accident to customers. He criticizes the PUC for allowing costs to be passed on. (HP)

"Group Would Ax N-Liability Act." The Pa. State Association of First Class Township Commissioners call for a repeal of the Price-Anderson Act limiting nuclear liability. The Price-Anderson Act is felt to be too limited if a massive TMI evacuation were necessary. (HP)

"Boro Wants Monitoring & Accurate Info. From Met Ed." The Middletown Borough Council passed a resolution which requires continuous monitoring of radiation at strategic locations in Middletown, a central source where up-to-the-minute TMI information can be received, and other modifications. (MPJ)

6/20/79 "Guidelines on Tonights Public Meeting Are Set."
Restrictions and time limitations on TMI related questions that might be asked by area residents at a Borough Council meeting are presented. (MPJ)

"Little But Words."
The article predicts the events of a meeting scheduled for 6/20 between PANE and the Middletown Borough Council. The meeting is predicted to be dominated by vehement opposition to nuclear energy and to the reopening of TMI. (MPJ)

6/21/79 "Washington Merry-Go-Round by Jack Anderson."
The 'unforeseen' at TMI was foreseen years ago. Anderson states that the federal government contention that problems at TMI were unforeseen is not true. He says the evidence indicates that atomic experts were worried about hydrogen gas problems at Unit II a decade ago. A report of the AEC on 9/5/69 is cited. Anderson demonstrates flaws in guidelines for nuclear reactor emergency core cooling systems. (HP)

"TMI Foes Generate Heat at Middletown Meeting."
In a meeting attended by more than 250 Middletown residents, the Borough Council was criticized for not passing a resolution calling for the closing of Unit I. (HP)

"Radiation Study Begun in 5-Mile Radius of TMI."
Forty thousand residents within a five mile radius of TMI are being surveyed

to find the effects of low level radiation on health. The CDC and the NIH are sponsoring a survey with coordination from the PDH. (HP and NYT)

6/21/79 "Plan Calls for NRC Member at N-Accident Site."
The NRC is considering a proposal calling for one of its members to assume immediate, overall control of a nuclear power plant in the event of an accident similar to TMI. (HP)

6/22/79 "'We'll Never Know' N-Dose."
The NRC says we will never know how much radiation escaped from the TMI plant because the levels exceeded the abilities of the plants to measure them. (HP)

"Pipe Flaw Found in Line to Unit 1."
A crack in a pipe that supplies emergency cooling water to the Unit I reactor was discovered. The unit is closed at this time. (HP; also 6/29/79 LIJ)

"Speaker Sees N-Power Hostility Dying."
Charles Robbins, former President of the AIF, calls for expanded commercial use of nuclear power. He believes the antagonism steming from TMI will subside. (HP)

"Incidents at N-Plants in Single Day."
Two incidents at Peach Bottom and one at TMI were reported to the NRC. Officials said that there was no cause for concern. (LIJ)

"City Bill $17,000 for TMI Battle."
In Lancaster's suit, legal fees so far are $17,000. (LIJ)

6/23/79 "Governor Urges NRC to Delay TMI Reopening."
Gov. Thornburgh urges the NRC to indefinitely postpone the reopening of the undamaged reactor at TMI. (LIJ and HP)

6/23/79 "Mrs. Denton Basks in the Glow of Her Husband's TMI Deeds."
Mrs. Denton and her children visit the area for a vacation. (LIJ)

6/27/79 "Transcripts of Presidents' Hearings on TMI Accident are Available."
Transcripts of public hearings held by the Kemeny Commission are available to the public. Descriptions of each transcript and its cost are given. (MPJ)

"President's Commission Sets More Hearings."
The article announces three more Kemeny Commission hearings. (MPJ)

"Citizens Demand More From Council."
Reported on are the results of a meeting between the public (mostly PANE members) and the Middletown Borough Council. Comments from members of both sides are given to show that the meeting was highly emotional and tense. (MPJ)

"Pane Gains Recruits; Urges Active Opposition to TMI."
PANE's motives, members, and activities are profiled. Two members, John Garver and Jim Hurst, comment on the need for more recruits to the group's cause, as there are only 50 members and more are needed to accomplish their goals. (MPJ)

"NRC Rep. Will be at Ertels' Town Meeting."
Victor Gilinsky of the NRC is to appear at a Middletown town meeting. Gilinsky's purpose is to answer questions and hear comments of area residents. (MPJ; also 6/29/79 HP)

6/28/79 "Middletown to Get TMI Resolution."
Middletown Borough Council is drawing up a resolution opposing reactivation of TMI until serious issues are resolved. The article states the Council has been

criticized for not taking a public stand. (HP)

6/29/79 "Meeting on TMI Set."
Highspire Borough Council will conduct a meeting on TMI for residents on 7/5/79. (HP)

6/30/79 "NRC Acts to Extend Shutdown for Unit 1."
The NRC Keeps Unit I shutdown until public hearings can be held on reactivation. It is speculated that the NRC is acting at the request of Gov. Thornburgh. GPU also announced a sale of mortgage bonds to other utility companies in order to weather a cash drain caused by the crisis. (HP)

7/3/79 "NRC Puts Lid on TMI Unit 1; Prepares a List."
The NRC officially ordered Unit I at TMI to be kept shut until further notice. In 30 days, the NRC will issue an order outlining conditions that have to be met before Unit I can be reopened. (HP, LIJ, and NYT)

"Council's Resolution Opposes Reopening of Unit 1."
The events of a Middletown Borough Council meeting are reported. A resolution passed reflects the strict opposition felt by the Council members in reference to the reactivation of the Unit I reactor at TMI. (HP and MPJ)

"Highspire Citizens Get Chance to Air TMI."
The article announces a hearing for Highspire residents as recommended by the TMI safety commission. (MPJ)

"NRC Rep. Disappoints Anti-Nukes."
Expression of the tension and emotion felt at a town meeting in Middletown is described. Cong. Ertel and the NRC's Victor Gilinsky fielded questions from

the public, and Gilinsky reported important NRC decisions concerning TMI to the audience. Several sample questions and answers are given, and the reporter determined that Gilinsky was evasive in answering. (MPJ)

7/3/79 "ACRS Subcommittee on TMI to Meet July 9th."
The article announces a technical meeting for the NRC subcommittee that monitors reactor safeguards. (MPJ)

7/4/79 "Area Man Seeks $20,000 of TMI Owners, Operators."
Calling the operation of TMI as "ultra-hazardous," a York Haven man filed a $20,000 damage suit against owners and operators of the plant in a US Middle District Court. (HP)

"TMI Center Will Reopen on Saturday."
The TMI Observatory Center will reopen. It will present a film explaining the 3/28 accident. (HP; also 7/5/79 LIJ and 7/8/79 NYT)

"TMI Detours Japanese."
Japanese legislators and labor leaders were not permitted to tour the TMI plant. The group is described as antinuclear. (HP)

7/6/79 "3 Panels Suspicious of TMI Core Cover Up."
Investigators are seeking to discover why the government was not immediately informed that the core of the TMI nuclear reactor began disintegrating within a few hours after the start of the accident. They ask whether there was a cover-up. LIJ and NYT)

"Health Research Census at TMI Nears Completion."
The PDH has contacted 85% of the households within 5 miles of TMI to study how health is affected by low level radiation. (HP)

7/6/79 "Panel Opens TMI Records to Anyone."
Under the Freedom of Information Act, the Kemeny Commission has opened its records to the public. (HP and NYT)

"HMC Will Study Accident."
The Robert Wood Johnson Foundation provided a grant to HMC for a three month study of human and professional behavior patterns, performance, institutional functioning, and decision making during the TMI crisis. (HP)

"Met Ed Rate Battle."
HEC has been denied their request in the Met Ed price increase. Hershey viewed the increase as primarily to help pay for Unit II. (HP)

"TMI Legalities Panel Grows."
A conference on Legal Consequences of Nuclear Accidents is to be held at Hershey, 7/27 and 7/28/79. Speakers and subjects for the panel are named. (HP)

7/9/79 "State TMI Panel Says U.S. Agency 'Uncooperative.'"
Pa. Commission on TMI officials charge that the NIMH has not cooperated on health related studies. (HP)

"TMI Movie, Unlike the Bubble, Is a Bust."
Patriot movie reviewer discusses a Met Ed movie on TMI. (HP)

7/10/79 "Thornburgh Gets N-Plant Plaudits from Governors."
Thornburgh appeared before the Committee on Public Protection at the National Governor's Conference to brief governors on what happened at TMI. (HP)

Editorial: "Remote Locations Are No Safer."
The editorial rejects the suggestion of placing nuclear plants in remote rural areas. The editorial claims this would

violate the 14th amendment on equal protection. The editorial contends that if plants are safe, they can be located in cities as well as rural areas. If plants are unsafe, they are unsafe anywhere. (HP)

7/10/79 "TMI Power Gap Pact Signed."
The GPU signs a $1 million contract with Ontario Hydro of Canada for replacement energy for stricken TMI. (HP and NYT)

"State Asks TMI Water Role."
The State of Pa. filed a motion to be a friend of the court regarding treatment and disposal of contaminated water at TMI. The State's position is described as neutral. (HP and LIJ)

"GPU Near Decision on Radioactive Water at TMI."
The GPU will soon announce its decision on how to dispose of radioactive water. (LIJ)

7/11/79 "Pro & Anti Nukes Talk at Highspire."
The article reports on a meeting in which four speakers confronted the views of Highspire residents on the subject of TMI. The speakers were State Sen. George Gekas, Dauphin County Comm. Stephen Reed, Dr. Leavitt, and Dr. Wm. Schoff. Each speaker expressed his views, then the public had its opportunity to speak. Most public views were antinuclear. (HP and MPJ)

"TMI Water Cleaning Eyed."
Chem-Nuclear Systems, Inc. agrees to design and operate a system that will purify radioactive water at TMI. The process uses an 'ion exchange column.' (HP and LIJ)

7/12/79 "City Unhappy with TMI Plan."
Lancaster City is unhappy that GPU plans to purchase the cleanup process before

an environmental assessment has been completed. (LIJ)

7/12/79 Editorial: "We Will Not Be Lulled." The editor alleges that people living under the ominous shadow of TMI do not want public utilities deciding how much radiation is safe for us, nor how much tritium or other radioactive substances we must accept in our drinking water. (LIJ)

7/13/79 "SVA Hears Status of TMI Lawsuit." Kohr updated members of SVA as to the current status of their lawsuit. (LIJ)

"NRC Will Keep TMI Unit 1 Shut Despite Firms Protest." The NRC ordered Unit II shut despite GPU claims that continued inactivity threatens the financial stability of GPU with bankruptcy. (LIJ and HP)

"GPU to Continue Reduced Dividend." GPU will continue its reduced dividend in the 3rd quarter. (NYT)

"TMI II Fire Put Out." A fire extinguished immediately by control room personnel caused minor damage to Unit II radiation monitoring equipment. (HP)

7/14/79 "Unit I Continued Shutdown Praised." Gov. Thornburgh praises a NRC decision to keep Unit I closed for up to 15 months until public hearings can be held. (HP)

"Excess Radiation at Middletown." About 100 workers at TMI were exposed in the last few weeks to radiation exceeding federal work limits. The exposures are not termed a health hazard. (NYT)

"City Charges NRC Abusing TMI Order." Lancaster City charges that the NRC decision giving Met Ed permission to purchase a demineralization process is in violation of an order requiring no action to be taken until an EIS is prepared. (LIJ)

7/16/79 "Senate to Weigh N-Plant Limits."
Congressional reaction to TMI will be tested by a new plan to place more restrictions on N plants. (HP)

7/17/79 "President Urges $140 Billion to Insure Energy Automony: Terms Nuclear Power Vital."
In a Kansas City speech, Pres. Carter says atomic energy must play an important role in the US energy future. (NYT)

"4-Year TMI Clean-Up Cost $400 Million."
Arnold of Met Ed said that the $400 million estimate of cleanup costs were developed by the BPC. The original cost of construction was $705 million. (HP, LIJ, and NYT)

"N-Evacuation Rule Approved."
The Senate voted to shutdown as of 6/1/80 all N plants for which a state has no federally approved emergency plans. It authorized $373 million to the NRC for programs including tightened control of nuclear industry. The TMI accident influenced these actions. (HP)

7/18/79 "Highspire Says Close N-Plant."
The Highspire Borough Council calls for permanent shutdown of TMI. Highspire is within a 5 mile radius of TMI. (HP)

"Goldsboro Man Files $20,000 Suit Over the TMI Mishap."
A Goldsboro man alleges owners of TMI operated the plant in a defective condition. He files a $20,000 lawsuit. (HP)

"Met Ed Bids U.S. Court Dismiss Suit."
Met Ed asks a federal court to dismiss a class action lawsuit stemming from the TMI accident. (HP)

7/18/79 "Schweiker's N-Plant Training Rule Gains." The Senate moved to tighten federal safety regulations at N plants. Sen. Schweiker makes a contribution to the bill which was dominated by a reaction to TMI. (HP)

"Clean Up to Start on Unit II." Cleanup plans intended for use in the removal of radioactive water in the TMI Unit II reactor are described. The water decontamination system known as Epicor II was intended for this purpose. Its functions and the disposal of its waste products are explained. (MPJ)

"Commission on TMI Agenda for July Hearings." The agenda for public hearings of the Kemeny Commission on TMI is made public. (MPJ)

"ACRS Advisory Group Sets Meeting on TMI." The article announces a technical meeting of the NRC's subcommittee in charge of reactor safeguards. The purpose of the meeting is to discuss TMI long term implications. (MPJ)

"City Delays Legal Action on TMI Water." Favorable talks with NRC officials leads the city of Lancaster to delay legal action. (LIJ)

"Legislators Ask No TMI Water Dump." Members of the Pa. House Select Committee on TMI urged owners to refrain from dumping decontaminated water into the Susquehanna River. (LIJ)

7/19/79 "SVA Seeking Injunction on Water Cleanup." The SVA is seeking an injunction preventing any cleanup action to be taken until an EIS is released. (LIJ)

7/19/79 "NRC Testing TMI Workers for Radiation." The NRC is testing 100 workers who were exposed to radiation from 6/30 to 7/4. (LIJ)

"TMI Unit I Sends Waste to S.C." Waste from Unit I left TMI enroute to S.C. The state earlier refused Unit II waste. (HP)

"B and W Brass Ignored Warning." Some B&W officials imply that the failure of Unit II might have been prevented. They tell the Kemeny Commission that lower level executives mulled over concerns prior to the crisis, but superiors did not heed their suggestions nor share the operators' concerns. (HP and NYT)

7/20/79 "B&W Executives Admit Neglecting Warnings." B&W executives told the Kemeny Commission that they failed to heed warnings that their reactors could be prone to the type of accident that shutdown TMI. (HP and NYT)

"Ways to Improve N-Safety Offered." A NRC task force recommended 23 immediate actions to improve N plants. These recommendations are based on lessons learned from TMI. Recommendations relate to reliability of operators, training of operators, system shutdown, and communications. (HP; also 7/22/79 NYT)

"Met Ed Relates Steps for Energy Management." Met Ed, following a 6/15/79 order by the PUC, filed a report on energy management. Met Ed must purchase power to replace power lost due to the TMI accident. The report also contains suggestions for reducing energy needs. (HP)

7/20/79 "Consolidation Ordered in Suits Against TMI."
Twelve lawsuits against TMI were ordered consolidated into a class action suit by a US magistrate. (HP)

"TMI Xenon Release 21 Times That Predicted."
A NRC report dated 5/9/79 and written by D.F. Bunch, Director of NRC's Program Support, stated that the Xenon release was 21 times greater during the accident than predicted. Despite this, officials state that the risk is minimal. (LIJ)

"Met Ed Files Report on Energy Conservation."
Energy conservation plans are filed with the PUC by Met Ed. (LIJ)

7/21/79 "NRC Working on Nuke Lawsuits Ground Rules."
The NRC is asking for public input to help determine whether the owners of TMI can use a certain defense approach in lawsuits stemming from the TMI nuclear accident. (LIJ)

"B&W Official 'Not Proud' on Precautions."
John MacMillan, a B&W official, told the Kemeny Commission that although he was 'not proud' of the firm's considerations on tougher safety standards prior to the N accident at TMI, he believed that B&W had adequately trained Met Ed personnel to handle the kinds of problems that occurred there. (HP)

7/23/79 "Utility Still Favors Dumping TMI Water in River."
GPU officials still believe their dumping plan is best. (LIJ)

"Views Vary on Effects of TMI."
Medical opinion varies widely over the extent of psychological damage caused by TMI nuclear accident, but most area

doctors agree that adults seem to be more vulnerable than children. (LIJ)

7/23/79 "Harrisburg Mayor Seeks Link With Hiroshima."
Mayor P. Doutrich of Harrisburg wants to establish a sister city relationship with Hiroshima. (NYT)

7/24/79 "Delay in N-Plants Urged."
The World Council of Churches' Conference on Faith, Science, and the Future recommended a moratorium on the construction of new nuclear plants. (HP)

"Documents Tell How to Break into N-Plant."
Barrier Penetration Database, an illustrated NRC report issued in 6/78, describes 32 physical security barriers commonly found at nuclear power plants. It details how to break these barriers. Other N plant related pamphlets are noted. (HP)

"Another Cloud Over TMI; Suits Piling Up."
Class and individual suits are filed against GPU and B&W. Damage is being sought for economic losses, threatened illness, and emotional distress. (LIJ)

7/25/79 "TMI Flushing 2 Towers with Susquehanna Water."
Periodic maintenance is being performed. (LIJ)

"NRC Shies Away From Dropping Promised TMI Hearing."
The NRC briefly considered not holding the public hearing that it promised prior to the startup of Unit I. (HP)

"TMI Spurs Oregon to Halt New Plants."
As a result of TMI, the state of Oregon has banned approval of new atomic plants in Oregon for at least 16 months. (HP)

7/25/79 "Phone Survey Probes Effect of TMI Accident."
The NRC is sponsoring a phone survey to determine the effect of TMI on Pennsylvanians. The survey is being conducted over a 90 mile radius by Chilton Research Services of Radnor. It will be used to evaluate social and economic costs. (HP)

"'Hot' Residue May Reach 100 Truckloads."
The contaminated residue from TMI may reach 100 truckloads. It is estimated 880,000 gallons of water are contaminated. (HP)

"Denied Access to Data, Citizens Panel Resigns."
The citizens panel formed to advise the Kemeny Commission met for the first time and resigned because it could not share in commission information. (HP)

"Government's N-Crisis Handling Hit."
Mayor Albert B. Wohlsen, Jr. and other Lancaster County officials criticize state and federal handling of the crisis before the Pa. House Select Committee on TMI. (HP)

"Met Ed Pres. Will Meet Residents."
Walter Creitz was to meet with Londonderry residents to candidly discuss TMI. Specific questions for discussion are listed. Also listed are provisions of a resolution drafted in response to TMI. (MPJ)

"Highspire 1st to Call for Shutdown."
Highspire Borough Council members voted for a complete shutdown of TMI. (MPJ)

"TMI Survey Results to be Discussed."
The article reports the agenda for a Lower Swatara Township Board of Commissioners meeting. Among other topics listed is the discussion of the township's TMI survey. (MPJ)

7/26/79 "Widoff Says Power Industry Might Bend Probe."
Mark Widoff, former state consumer advocate and a member of the defunct citizens advisory committee to the Kemeny Commission, warned that N power industry interests might unduly influence the commission's investigation. (HP)

"Minnich Rips State Handling of TMI Crisis."
Dauphin County Commissioner, John E. Minnich, related the difficulties he had in obtaining information from the state during the TMI crisis. (HP)

"NRC Eyes Releasing TMI Radioactive Gas."
The NRC is eyeing releasing Krypton 85 into the atmosphere. (LIJ)

"Why is Vapor Coming Out of Towers at TMI?"
Met Ed officials explain the presence of vapor from the TMI towers as an end product of water circulation which is needed even though electricity is not being produced. (LIJ)

7/27/79 "Elderly Will Review TMI-Related Plans."
Senior citizen groups are among consumer groups who plan to review the energy conservation plans of Met Ed. The plans are part of a PUC ruling to minimize the cost of power normally supplied by TMI. (HP)

"NRC Competence Doubted."
The House Government Operations Committee is reported ready to release a document stating that the NRC is unprepared to handle emergencies like TMI. (HP)

"NRC Advised Nuke Plants to Check Welds."
The NRC is advising N plants similar to TMI to be checked for weld cracks in the emergency cooling system. The weld cracks were probably caused by the stress of the accident. (LIJ)

7/28/79 "Customers Bear TMI Bill."
John Hoffman, a consumer lawyer, states ratepayers had nothing to do with the TMI accident, but they will pay the bill. The burden being payed by consumers is four times that of stockholders. (HP)

"N-Accident Planning 'Chase' Seen."
Despite lessons from TMI, the House Government Operations Committee feels emergency planning is chaotic and inadequate. They blame the NRC. (HP and NYT)

"Halving of N-Plant Building Time is Pursued."
The Carter administration is set to revive a proposal to cut time to build a N plant from 12 to 6 years. The proposal will wait until Carter receives a final report from his TMI commission. (HP)

7/29/79 "After 4 Months, Three Mile Island Plant Isn't Cool Enough for a Cleanup."
The article assesses the status of TMI four months after the accident. (NYT)

7/31/79 "Met Ed Water Discharge into River Stopped by NRC."
A federal inspector stopped the discharge of water from Unit I into the Susquehanna River because it was not analyzed for Strontium 90. (HP and LIJ)

"Walker Asks NRC Lift TMI License."
Cong. Walker asks the NRC to immediately and permanently lift Met Ed's license to operate. (LIJ)

8/1/79 Editorial: "Ignoring the Lessons of TMI."
The editorial states that the people of Lancaster County can never again have peace of mind as long as the Susquehanna River, the source of drinking water for thousands of homes, is being used as a nuclear sewer. (LIJ)

8/1/79 "Mental Angle Stumps NRC as TMI Topic."
The NRC is said to be uncertain whether the mental health effects of Unit II should be a factor in reactivating Unit I. (HP)

"GPU Earnings Down 26.9%."
GPU earnings for the 2nd quarter, 1979, dropped 26.9%. (HP and NYT)

Editorial: "Met Ed Gives Chutzpah New Meaning."
The editorial states that a Hebrew term, chutzpah, is said to be given a new meaning by Met Ed due to their act of dumping radioactive garbage into the Susquehanna River. (HP)

"Unit 1 'Dumping' Upheld, Hit."
The article describes dumping of 4,000 gallons of processed water into the Susquehanna River. Pro and con viewpoints are presented. (HP)

"TMI and You."
An advertisement from Met Ed/GPU on TMI is presented. The ad announces plans for an information series and also regrets that a better job of communication was not done. (HP)

"Anti-Nuke Protests Pay Off in Township."
The Lower Swatara Township voted unanimously to permanently close TMI. In making their decision, the council members first consulted the public by means of a survey. The survey results are given. (MPJ)

"NRC Invites Public to Assist in Determination of TMI Mishap."
The NRC is inviting the public to submit information which will assist the Commission in determining whether the waivers of the defense provisions of the Price-Anderson Act should be applied in lawsuits involving the 3/28 accident at TMI.

Details of the Act's implications and of the importance of public response to this NRC request are given. (MPJ)

8/1/79 "23 Changes Called for to Improve Nuke Safety."
A NRC task force study of the TMI mishap made 23 recommendations for the improvement of public protection. (MPJ)

"Radiation & Health Conference to Feature Top Nuclear Power Experts."
The article announces a two day conference entitled 'Radiation & Health.' An extensive agenda for the conference is given, and a forum entitled 'How We Coped' reviews the events of the TMI mishap in terms of stress. (MPJ)

8/2/79 "State Can Join TMI Dump Suit."
Pa. will be allowed to join the suit to prevent dumping of water from Unit II. A related story describes GPU's plan to realign management and staff at TMI. (HP)

"New Cumberland OK's TMI Paper."
New Cumberland Borough Council passes a resolution aimed at keeping TMI closed until safety questions can be answered. (HP)

"Veto of Water Study Expected from NRC."
Lawyers for the SVA predict that the NRC will decide next week that an EIS before cleanup is not needed. (LIJ)

8/2/79 "GPU Hopes to Change TMI Discharge Views."
Dieckamp feels that area residents have a 'psychological' fear of dumping. He hopes to present facts to change their views. (LIJ)

8/3/79 "Tests at TMI OK: Dumping to Continue."
Even though tests of last week's discharge are incomplete, NRC officials have given Met Ed permission to continue dumping. (HP and LIJ)

8/3/79 "TMI Mishap Preventable, NRC Believes."
NRC staff (Office of Inspection and Enforcement) believe the accident at TMI would have been prevented if plant operators had let safety equipment function as it was designed to do. (LIJ and NYT)

"Nuclear Water Called Safe."
Met Ed says the level of radioactive Strontium 90 in 4,000 gallons of water dumped into the river was 10% of the concentrations allowed by the federal government. (NYT)

"Tulsa Hearings Restart Asked on Nuke Plant."
The accident at TMI delays the construction of Black Fox energy plant in Oklahoma. Hearings on the Oklahoma plant will reopen in hopes of answering TMI Questions. (HP)

"GPU Head Hits NRC Criticism."
Dieckamp charges that the NRC report on TMI unfairly criticizes Met Ed. B&W says the report properly puts the blame on control room operators. (HP)

"N-Power Lid Still Rated Possible."
Pa. officials concede they were ill prepared for TMI. Kemeny expresses the belief that Pres. Carter will pursue all of the group's recommendations including a ban on nuclear power generation. (HP)

8/4/79 "Group Sets Conditions for TMI Unit."
The HAAANP asks the Derry Township Board of Supervisors to support a resolution opposing the reopening of TMI. Lists conditions for reopening. (HP)

8/7/79 "State House Committee Votes TMI Nuclear Ban."
Pa. House of Representatives Professional Licensure Committee approved a bill to ban N generation on TMI, but leave facilities open to any other type of electrical power generation. (HP; also 8/8/79 LIJ)

8/8/79 "Hearings on TMI's Unit 1 Restart Lined Up."
The NRC announces that public hearings on the restart of Unit I will probably start in 2/80. (HP)

"Amish in Dark on TMI Crisis."
The Amish were in the dark in terms of communication on TMI. The Amish population in the area is approximately 30,000 including 15,000 Old Order Amish in Lancaster County. Their main source of information is a weekly paper. It is estimated that one week passed before all the Amish knew of the problems. (HP)

Editorial: "Terminating TMI."
The editorial calls for the state legislature to openly face the N risk to the stability and well-being of the state. It disagrees with the suggestion made by State Rep. Eugene R. Geesey that TMI be converted to a fossil-fuel generating facility. (HP)

"Unit 2 Waste on Way West."
The first waste from Unit II is on its way to Washington State. Other TMI developments were discussed including a plant worker and ambulance driver who were contaminated by materials from Unit II and the beginnings of the attempt to sample water inside Unit II. (HP)

"NRC Issues Conditions of Unit 1 Reactor Start."
Ten short range and five long range conditions were listed by the NRC. Met Ed has 20 days to respond. (HP; also 8/9/79 LIJ)

8/9/79 "TMI Boss: Coal Not Out of Question."
A study was launched by GPU to examine the feasibility of converting TMI to a coal-fired plant. (LIJ)

8/9/79 "N-Plants Viewed as U.S. Answer to OPEC." L. Rukeyser, national financial advisor, suggests that if we close our eyes to the potential of nuclear energy, OPEC will profit. He states that a TMI report from the Kemeny Commission may be reassuring. (HP)

"TMI N-Waste Crossing Midwest." The article discusses the progress of a truck with a load of N waste as it moves across country to Washington State. (HP)

"Guardmen's Ranks Thin in N-Crisis." Many national guardsmen opted to leave the area with families instead of remaining in the area in case mass evacuation was needed. (HP)

8/10/79 "Revised N-Area Evac. Plans Unveiled." York County EMA revised its plans for evacuation of residents from 5, 10, and 20 miles around TMI and the Peach Bottom N power station. (HP)

Editorial: "No Connection." The editorial criticizes the attempts to compare the TMI accident with the bombing of Hiroshima. (HP)

"NRC Sets Agenda for TMI-1 Hearings." The agenda for TMI hearings in Harrisburg relative to Unit I is set. A proposed schedule is presented. (HP)

"2 From Area of TMI to Train in N-Navy." Two local men who experienced TMI express their views on N power. Both will enter the Navy's N power schools. (HP)

8/11/79 "Post-TMI Releases PR Type." Met Ed has launched a PR campaign to convince the people of Central Pa. that N power is safe. (LIJ)

8/11/79 "State Will Monitor Nuke Pregnancies."
Dr. George Tokuhata, Director of the BER of PDH, announced that the state will study effects of low level radiation and the fear factor on pregnancies. (HP, LIJ, and NYT)

"TMI Start-Up is Resisted."
Londonderry Township, where TMI is located, approved a resolution opposing reactivation of Units I and II until a series of conditions are met. (HP)

Editorial: "Overdose of Stress."
The editorial calls stress one of the most significant health threatening aspects of TMI. The NRC is uncertain how stress relates to its deliberations on TMI. The editorial calls for consideration of stress and its effects on the area. (HP)

8/13/79 "Nuclear Foes to Ask Help of Derry Township."
The HAAANP will ask the Derry Township Board of Supervisors to endorse a resolution opposing the reopening of Units I and II. (HP)

"Seminar Set For Educators on 'Radiation.'"
The Capitol Campus of PSU will host a conference for area educators on radiation, nuclear energy, and radiological health. Dieckamp will discuss 'Impact of TMI-2 Type Accident Upon a Utility.' (HP; also 8/15/79 MPJ)

"TMI Panel Hears Everything but Names."
No local officials knew who had the authority to order evacuation. (HP)

8/14/79 "TMI Tells How it will Release Radioactive Gas."
Radioactive gas trapped at TMI will be vented into the atmosphere over a 30 day period sometime this fall. (HP, LIJ, and NYT)

8/14/79 "PUC Urged to Cut TMI Rates."
State Rep. Rhodes urged the PUC to remove the cost factors associated with TMI from consideration in the existing rate, thereby lowering it. (LIJ)

8/15/79 Met Ed Ad: "TMI and You-Radiation."
The ad quotes an Ad Hoc Population Dose Assessment Group, a federal government project, stating that only one person is likely to die due to the radiation released from TMI. Therefore, it is safe. (LIJ and HP)

"NRC Favors Processing of Water at TMI Unit 2."
The NRC recommended that 280,000 gallons of moderately radioactive water stored in tanks at Unit II be processed by the Epicor II system. Public comment is invited. (HP, LIJ, and NYT)

"Doings at TMI Produce Dissent at Lancaster, Here."
The SVA objected to the NRC report that endorses the Epicor system to remove radioactive water from TMI. In Harrisburg, other protestors urged Gov. Thornburgh to stop the venting of the gas at TMI and to block the dumping of the water into the Susquehanna. (HP)

"Parent of B&W Bouys 'Nuclear.'"
James E. Cunningham, Chief Executive of B&W, urges the country to continue its atomic power research program. (HP)

"Londonderry Residents Express TMI Views to Met Ed Executives."
A meeting held between Walter Creitz and Londonderry residents is reviewed. The purpose of the meeting was to reassure the public that the cleanup of Unit II and the reopening of Unit I would not cause a health threat. Comments made by various members of the public are given. (MPJ)

8/15/79 "TMI Waste Will Go To Washington."
The article explains the methods of solid waste disposal used by TMI. (MPJ)

"Thornburgh to Address TMI Commission."
A schedule of public hearings to be held by the Kemeny Commission is listed. (MPJ)

8/16/79 "Consumer Advocate Limits Role on TMI Question."
State consumer advocate, Walter W. Cohen, filed petitions to intervene in federal hearings on the startup of Unit I. He hopes proceedings do not become a review of the future of N power in the US. He is neither pro nor con on N power, but only wants to see the economic interests of power customers protected. (HP and LIJ)

"Radiation Risk is Downgraded."
The NRC says that tests of processed radioactive water dumped into the Susquehanna River from Unit I revealed less radiation than previously cited by the operator. (HP and LIJ)

"Legislator Tell SVA of Pending N-Debate."
State Rep. Brown informed SVA members of a pending legislative debate on the bill to permanently keep TMI shut. (LIJ)

8/17/79 "Worker at TMI Suffers Slight Radiation Dose."
An employee involved in recovery operations at TMI touched the harness of a discarded respirator and received a dose of 1 millirem on his finger. (HP)

"Key Damage Test Set at TMI."
Water will be drawn from the flooded basement at Unit II to give an indication of damage to the reactor. (HP)

8/17/79 Editorial: "TMI Consumers."
The editorial expresses the feeling that consumer advocate Cohen may not adequately represent consumer interest concerning TMI to the NRC. (HP)

"PUC Needs Time to Cull TMI Data."
The PUC needs at least six months before it can submit its findings from an investigation of the management practices of the firms involved with TMI. The probe focuses on construction and operation. (HP and LIJ)

"4 Groups Demand TMI Shutdown."
The four groups, calling for a permanent shutdown of TMI, are: 1) Friends of the Earth-Philadelphia 2) The Nuclear Watch Committee 3) Pa. Chapter of the Sierra Club 4) Eastern Pa. Chapter of the Sierra Club. (LIJ)

8/20/79 "State Drops N-Waste Site Plan."
The psychological effect of TMI has forced Pa. to abandon its plans for a low level nuclear waste site. (HP)

Editorial: "N-Mood."
The editorial states that experts have devised plans to remove water and gas from TMI, but the public is skeptical. The editorial advocates that technology must conform to the wishes of the people, not vice versa. (HP)

8/21/79 "Beeper Sounds False Alarm."
A telephone beeper, connected to a radiation monitoring device on a water discharge pipe at TMI, malfunctioned and went off three times in an eleven hour period. (HP)

"TMI Hearings to Resume."
Gov. Thornburgh will appear before the Kemeny Commission. The State House Select Committee on TMI will also resume hearings. (HP)

8/21/79 "Property Keeps Pre-TMI Values."
According to realtors and York and Dauphin County assessment records, the TMI accident had no apparent effect on property values. (HP and NYT)

8/22/79 "Panel Skeptical of N-Accident Evac. Plan."
State legislators on the State House Select Committee on TMI questioned if the state had a workable evacuation plan in the event of a reocurrence of a N accident at any of the commonwealth's reactor facilities. (HP)

"Polite Silence Greets Met Ed President."
Walter Creitz spoke to citizens in Londonderry Township. Pro and con N power statements were made. The article also discusses the pressure citizens have placed on local officials for resolutions on TMI. (HP)

"Royalton Seeks TMI Reaction."
Royalton Borough is looking at ways of determining public reaction to TMI. A resolution was passed calling for adequate monitoring and warning devices at plant expense. (HP)

"Swatara Resolution on TMI Policy Due."
Swatara Township will consider a resolution on TMI. Local officials report some pressure, but not as intense as in some other TMI area communities. (HP)

"TMI Query Stumps Thornburgh."
Gov. Thornburgh appeared before the Kemeny Commission. He discussed his role in the crisis. He was reported as being unprepared on questions of National Guard readiness during the crisis. There were reports that the ranks were thin. (HP)

8/22/79 "Radiation Monitors Finally to be Placed in Boro by Met Ed."
Met Ed informs Middletown Mayor Reid that radiation monitors will be placed in the Borough. Previously, PANE demanded a status report. (MPJ)

"Agenda Set for Commission Hearings on TMI."
An agenda has been set for the 8/21-23 public hearings on TMI by the Kemeny Commission. (MPJ)

"NRC Issues Environmental Assessment of Decontaminating TMI Waste Water."
The NRC is releasing for public comment an environmental assessment of specially built equipment to decontaminate intermediate radioactive waste water from TMI. The Epicor II system and alternatives are discussed. (MPJ)

"Mayor Writes 3 Papers Expressing TMI Concern."
In exactly worded letters to the *NYT*, *Washington Post*, and the *Wall Street Journal*, Mayor Wohlsen explains why concern about TMI has not ended. The primary issue is safe water. (LIJ)

8/23/79 "Issued 'Pre-TMI' Warning, Inspector Says."
James Creswell, an inspector in the NRC's Chicago office, says he tried to alert the NRC to safety procedures that might have helped Unit II operators to prevent the accident. (HP and NYT)

"Radiation Touches Pants."
A worker taking part in the TMI cleanup wore home a pair of trousers contaminated with 2 millirems of radiation per hour. A plant detection device failed to find the contamination. (HP and LIJ)

8/23/79 Editorial: "Lid of Secrecy."
The editorial alleges that Switzerland withheld information from citizens on an accident there that resembled TMI. It calls for open disclosure of accidents. Disclosure may help prevent other accidents, but, the editor feels that disclosure of the Swiss accident would not have made a difference at TMI, since the NRC did not react to other warnings. (HP)

"TMI Memo to Guard Deplored."
Kemeny commented negatively on Pa. Adjutant General Richard Scott's advice to national guardsmen that they wouldn't be exposed to radiation danger if deployed for a TMI mass evacuation. (HP)

"TMI Breeds Absenteeism"
There are high rates of absenteeism on the Pa. House Select Committee investigating TMI. The Committee hears reports that some tourists stay away and business losses are projected. (HP)

8/24/79 "State Asks for Voice on Unit I."
Gov. Thornburgh petitions the NRC to grant 'interested state' status to Pa. when hearings begin on the reopening of Unit I. (HP)

"TMI Panel's Wrath Spurs Denton Shift."
Denton reversed a previous decision to end the moratorium on new power plant licensing and construction. The decision is traced to the Kemeny Commission. (HP and NYT)

"Thornburgh on TMI: I Won't Take a Stand Where I'm Not An Expert."
Gov. Thornburgh avoided specific responses to various proposals for the cleaning up of TMI. (LIJ)

8/25/79 "Protest Set on Venting on TMI."
TMIA will protest further releases of radioactivity at TMI. (HP)

8/26/79 "Water Samples Secured at Crippled Atom Plants."
The first water samples of radioactive water are drawn at TMI. (NYT)

8/27/79 "300 at Anti-TMI Rally on Steps of State Capitol."
A TMIA rally draws a crowd to the state capitol; SVA also joins the protest. (HP)

8/28/79 "TMI-1 Reopening Passes Vote."
Derry Township supervisors passed a resolution calling for the reopening of Unit I. (HP)

8/29/79 "NRC Plays Down River Impact of Berwick Nuke Plant."
The NRC says the proposed Berwick power plant will have no significant environmental impact. The plant is 70 miles above Harrisburg on the Susquehanna River. Berwick reactors are the boiling type as opposed to the pressurized reactors found at TMI. (HP)

"3 TMI Workers May Have Been Contaminated."
Three TMI workers received small radiation doses in two separate incidents. Two cases involved a radioactive water leak. (HP and LIJ; also 8/31/79 NYT)

"Subcommittee Studying TMI Schedules Sept. 5th Meeting."
A subcommittee of the NRC, ACRS, will hold a technical meeting in Washington. Rules for participation are outlined. (MPJ)

8/30/79 "Met Ed's Chief, Creitz, Resigns."
Walter M. Creitz has resigned five months after the TMI accident. He will continue on a special assignment at GPU. (HP and LIJ)

"TMI Opponents Picket NRC."
PANE picketed the NRC headquarters in D.C. (HP)

"Newberry Board Chief Wants a Better Warning System."
The chairman of the Newberry Township Board of Supervisors told a subcommittee of the Governor's Commission on TMI that a more efficient warning system must be developed in the event of a future N accident. He criticizes the reliance on fire sirens, saying they can't be heard by everyone. (HP)

8/31/79 "Acted Properly in TMI Support, Derry Township Says."
The HAAANP questioned the actions of Derry Township supervisors who voted to support the reopening of Unit I. The HAAANP asked that both Units I and II not be reopened until fail safe protections were found to protect the building. (HP)

"Panel Summons Leading Officials of Met Ed, GPU."
The Pa. House Select Committee on TMI summoned Met Ed and GPU officials to testify next week on their actions relating to the accident. (HP)

"6 at TMI Get Radiation Overdoses."
In the second incident of the week, six TMI workers suffered radiation exposure. Men were repairing a leaking valve in the building between the Unit I and II reactor containment building. The dosage is said to be a maximum body dose of 1 rem. (HP and LIJ)

9/5/79 "GPU Official Named Interim Met Head."
Floyd Smith is named interim head of Met Ed. (LIJ)

9/6/79 "Met Ed Employee Tries to Reassure Borough."
Robert Arnold attended a Royalton Borough Council meeting. He attempted to convince listeners that the safety of people in the area is a top concern of Met Ed. (MPJ and HP)

"TMI Happenings."
The article contains a calendar listing of TMI related meetings, discussions, and dates. (MPJ)

"Public Given Extension to Testify at Hearings."
The deadline is extended for the public to petition for full participation in a public hearing on reopening Unit I. (MPJ and HP)

"NRC Advisory to Meet on Reactor Safeguards."
The implications of the TMI accident and underlying causes will be on the agenda of an NRC meeting. The ACRS is now meeting in D.C. (MPJ)

"Report on TMI Workers."
Three masonary workers received small radiation doses on their knees. Six other workers received exposures earlier in the week. (MPJ)

"Swat, Steelton Meet on TMI."
The article lists scheduled meeting and speakers for the Swatara Township-Steelton Community Group of TMIA. (MPJ)

"Economic Fallout Up From Study."
Four federal agencies to study TMI effects at the cost of $686,000. (HP)

9/7/79 "NRC Sidesteps Immediate Decision on New Power Plants."
The NRC announces it will not decide on construction or operation of new power plants before the Kemeny panel makes its recommendations on the future of N energy in October. The NRC agreed to allow Harold Denton to order TMI-related safety requirements for 70 operating plants. (HP and NYT)

"Bishop Asks N-Plant Development Halt."
Bishop Joseph Daley of the Catholic Diocese of Harrisburg called for a temporary halt in the development of N plants. The plea is issued in terms of respect for human life. (HP)

"Nuclear Waste Disposal Woes Aired."
Testimony before Pa. House Select Committee indicated that failure of Pa. to search for a radioactive waste disposal site could lead to the loss of the right to dispose of waste elsewhere. (HP)

"PUC Asks Status in TMI Hearings."
The PUC asks for intervenor status in federal hearings on the reopening of Unit I. The PUC sees the proceedings as having a possible effect on Pa. ratepayers. (HP)

"Limited Evac. on TMI Questioned."
A NRC official says advising pregnant women and preschool children to leave the TMI area during the emergency may have been a mistake. Communication problems were cited as leading to the order. (HP)

"Geesey Assails July Dumping of TMI Water."
State Rep. Eugene Geesey of York said that the dumping of 4,000 gallons of waste water into the Susquehanna in July was a case of disregard for NRC recommendations. (HP)

9/8/79 "Radiation Incident Told."
A contractor left the TMI plant with a small amount of radiation on his work clothing. The incident was detected the next day. Three others also were contaminated, but it was discovered before they left the plant. (HP)

"TMI Parley Set."
TMIA will hold a hearing to assess Swatara Township and Steelton residents' opinions on the accident at TMI. (HP)

9/11/79 "GPU Engineer Sees Coal, N-Power as Viable Fuels."
A GPU engineer feels that the negative aspects of N power are far less than those of fossil fuels. He believes that TMI had a serious impact on the use of N power, but feels it won't last long. (HP)

Editorial: "N-Wastes."
The editorial supports statements of Robert C. Arnold that Pa. should consider an in-state disposal site for N waste. The state must face the problems of N power with safe disposal of waste products high on the list of problems. (HP)

"Funding Priorities Criticized."
The director of Dauphin County's Office of Emergency Preparedness criticized awarding federal grants totaling $686,000 to study the effects of TMI on tourism and business in the area. He cannot get money for emergency radio equipment. (HP)

"Group Seeks N-Accident Status."
TMIA has petitioned the NRC to declare the TMI accident an 'extraordinary N occurrence.' Under the Price-Anderson Act, such a designation would eliminate the need of persons who file lawsuits in connection with the accident to prove negligency on the part of Met Ed and others. (HP)

9/11/79 "No Nukes Criticize N-License."
The SVA charges that the NRC did not consider cause and effects of the TMI accident in giving PP&L license to operate a N plant in Berwick. (HP)

9/12/79 "Class 9 Action Buoys Critics of N-Power."
The designation of the accident at TMI as a class 9 accident will have an impact on the licensing of future plants. It is the first time the government has conceded to an accident of such magnitude. Future plants must be able to withstand such accidents. (HP)

"N-Utility Stocks Rebound, But."
Securities of N utility companies have rebounded from the TMI accident, but many investors remain wary of the future of N power. (HP)

"Alliance Fights to Keep TMI Shut."
The HAAANP begins a drive to keep TMI Unit I shut. (HP)

"Royalton Questionnaire Solicits Views on TMI."
Royalton Borough Council has invited Met Ed and NRC spokespersons to their meetings. A survey of area residents has also been mailed. Royalton is one of the few local communities not to show strong opposition to TMI. (HP)

"PANE Retains Counsel and Incorporates."
PANE has retained legal counsel and will incorporate. PANE will seek to intervene before the NRC on the subject of psychological trauma. (MPJ)

"TMI Incident Termed a 'Maximum Accident.'"
Dr. M. Kaku, a City College of New York N Physicist, labeled TMI a maximum accident. He spoke at F&M under the auspices of the SVA. (LIJ)

9/12/79 "Marietta Calls for Immediate TMI Shut-down."
Marietta Borough Council passed a resolution calling for the immediate closing of TMI. (LIJ)

9/13/79 "N-Core: Less Damage?"
Preliminary findings from the analysis of contaminated water taken from Unit II indicates less damage to the core than originally estimated. Estimates of damage to the core are revised from 50% down to 40%. (HP)

"River Study Set on Radioactivity."
The DER has commissioned a study to investigate whether radioactivity from TMI is accumulating in the Susquehanna River and its food elements. (HP)

"Official of TMI Resigns."
James L. Seelinger, Superintendent of Unit I reactor, has resigned. (HP; also 9/19/79 MPJ)

"Swatara Housing Project Opposed."
Residents protest the reactivation of the TMI plant at a Swatara Township Commissioners meeting. They pressured the Council to pass an anti-TMI resolution. (HP)

"Pennsylvania Nuclear Plant Seen in Use Again by '84."
Denton says TMI could be in use again by 1984 if it can meet stringent safety standards. His testimony was before the Pa. House Panel. (NYT; also 9/14/79 HP)

9/14/79 "TMI Insurance Claims Reviewed."
Insurance claims filed in the aftermath of TMI show little evidence of either personal injury or off-site property damage. Details testimony before the State House Committee on TMI. (HP)

9/14/79 "N-Levels Lower Than Predicted."
Met Ed announces that preliminary analysis of water within Unit II indicates less radioactive material than expected. (HP)

Editorial: "TMI Releases."
The editorial lauds Denton's statement that an order will be issued prohibiting the release of the radioactive elements at TMI. The editorial urges the NRC to declare that no contaminants can be released into water or air. (HP and LIJ)

"TMI Did Not Change Nuclear Experts 'Mind.'"
Dr. N.C. Rasmussen stated that N power is safe. He said that, "If we wait for a no-risk solution to the energy problem, I suspect we'll get no solution. And that may be the biggest risk of all." (LIJ)

"Mayor Hits TMI Treatment Study."
Mayor Wohlsen complains that the NRC's study is grossly inadequate because the NRC officials are too closely allied with Met Ed. He states that there is a need for an independent audit. (LIJ)

9/15/79 "Gekas: Keep TMI Closed."
State Sen. George Gekas asks the Kemeny Commission to recommend that TMI will never open again as a N facility. (HP)

"Berwick Plant Impact Asked by DER."
The DER asks the NRC to review parts of the EIS on the Berwick Plant of PP&L. The DER wants the statement to include an analysis of the impact of TMI lessons for the Berwick plant. (HP)

"Doctors TMI Diagnosis: Massive Dose of Stress."
Drs. J. Kalen and E. Martin are psychiatry professors at HMC. Based on interviews of 200 'neighbors' of the TMI plant,

they conclude that 'massive collective stress' was experienced. (HP)

9/17/79 "Nuclear Unit Urged to Weigh Accidents in Considering Sites."
O'Leary, formerly of DOE and AEC, warned Pres. Carter's campaign policy committee in 1976 that frequency of serious and potentially catastrophic nuclear incidents support conclusions that a major disaster will occur at a nuclear facility. (NYT)

9/18/79 "Borough Spurs Close TMI Bid."
State Rep. Stephen R. Reed plans to introduce legislation that will demand TMI be closed as a N facility. Credit for inspiration for a series of TMI related legislation given to Middletown Borough Council. (HP; also 9/19/79 MPJ)

9/19/79 Editorial: "N-Watchdog."
The editorial urges Pa. to take a more active interest in the five N plants that are planned to go into operation in the state in the next eight years. The DER is urged to play an important role as watchdog. (HP)

"SVA Finds NRC Report Unacceptable."
The SVA has issued its own report criticizing the NRC analysis of the safety of Met Ed's water treatment plan. (LIJ)

Met Ed Ad: "An Important Message for Met Ed Customers."
The ad advocates energy conservation. (LIJ)

"ACS Meeting to be Devoted to TMI."
The ACS Southeastern Pa. section will meet at PSU Capitol Campus. The subject is 'TMI-Where Have We Been and Where Are We Going.' (MPJ)

9/19/79 "Water Being Monitored Near TMI Plant." The Academy of Natural Sciences is monitoring the Susquehanna River near TMI for presence of radiation in the algae, fish, sediment, and clams. (MPJ)

9/21/79 "PUC to Scrutinize TMI-1 Cost Burden." The PUC has directed Met Ed to show why its customers should continue to pay costs associated with the operation of TMI Unit I after 1/1/80. (HP and LIJ)

"N-Licensing Board Sets Schedule." The article contains the tentative schedule for hearings set by the NRC on the reactivating of Unit I. (HP)

"Pro-Anti-Nukes Square Off at Londonderry Elementary." Sixth graders stage a debate on N power at Londonderry Elementary School. (HP)

"Close TMI, Columbia Demands!" Columbia Council passes a resolution opposing the restart of TMI. (LIJ)

9/22/79 "Met Ed Eyes Rate Hike Bid." Met Ed needs 4-6% rate hike to cope with costs stemming from the TMI accident. (HP and LIJ)

"Editor Says Met Ed Fumbled with Facts." Saul Kohler, Executive Director of Patriot-News Co., told a Pa. House Committee that the failure of Met Ed to level with news media hindered their presenting facts on the TMI case. (HP)

9/23 & 9/24/79 "Three Mile Island Accident: A Cloud Over Atom Power." This is a two part article on the TMI accident discussing political, financial, and technical developments that have made the accident the most serious challenge to the nation's civilian N power program in its 25 year history. (NYT)

9/24/79 "Anti-Nuclear Fight Can Be Won, Fonda Asserts."
Actress Jane Fonda and political activist husband Tom Hayden told 1,000 persons at a TMIA rally that the anti-nuclear fight can be won. (HP; also 9/26/79 MPJ)

"TMI Accident Cost $18.2 Million in Lost Wages, Expenses."
The results of an Economic Loss Survey were released by the NRC. (LIJ)

9/25/79 "Manbeck Writes Voters His Reaction to TMI."
State Sen. Manbeck supports five proposals to help operate TMI safely. He opposes shutdown of TMI. (LIJ)

9/26/79 "TMI Status Briefing Set by Met Ed."
A Met Ed official will hold a briefing on the future of TMI. (HP and MPJ)

"Met Ed Offers Revised Crisis Plan."
The NRC Emergency Planning Task Force is meeting in Middletown. Met Ed presented a draft outline of an emergency plan for the Unit I reactor. The plan includes organizational changes, communication improvements, and plant equipment modifications resulting from the TMI accident. (HP)

"Steelton Council Asked to Adopt Resolution Calling for End to TMI."
A petition to adopt a resolution calling for the permanent closing of TMI has been presented to the Steelton Borough Council. (HP)

"Joe Kachnoski and TMI: A Look From Shippensburg."
This is an interview with a Shippensburg State College football player who comes from Middletown. He discusses TMI. (HP)

9/26/79 "Subcommittee on TMI to Meet."
The NRC ACRS will meet in D.C. on 10/2 to discuss the response of the reactor manufacturers to the NRC bulletin. (HP)

"TMI 'Replay' Scheduled on Paper, Plastic Stage."
The NRC has hired Easer Co. of Arlington, Va. to simulate the TMI accident. (LIJ)

9/27/79 "Financial Woes Taking Toll on Owners of TMI."
The story outlines the financial problems facing GPU and its customers as a result of TMI. (LIJ and HP)

"Panelists May Urge NRC Split."
It is reported that one third of the members of the Kemeny Commission want the NRC stripped of its licensing and emergency planning power. (HP)

9/28/79 "Evac. Scope Called Confusing."
The NRC expanded TMI evacuation zone when fears of an explosion and meltdown existed. Previously, the evacuation zone was based on population density and was about a five mile radius. The new zone was extended from five to twenty miles in radius. The public is said to have been confused over the scope of the zone. (HP)

"TMI Will Be Missed if Winter Tough, PP&L Head Says."
A PP&L spokesman predicts brownouts and blackouts if the winter is excessively cold. Anticipated problems are due to the closing of the TMI facility. (HP)

"No Danger of TMI Core Melt Down, DER Head Says."
Clifford Jones states that at no time was there any danger of a hydrogen explosion or meltdown. (HP)

9/29/79 "TMI Leak Hinders Cleanup."
US Senate investigators for the NRC state that a radiation leak hinders cleanup. Mayor Wohlsen and the SVA dispute the urgency to start cleanup. Adequate storage space is reported available until safe cleanup is approved. (LIJ and HP)

"Discharge Intent Denied."
Arnold admits that the flow of radioactive water could fill Unit II tanks in 40 days. He says Met Ed will not be forced to let the water go into the river. (HP)

"Radiation Illness Medicine Pursued."
MacLeod urged the federal government to stockpile potassium iodide in the areas surrounding N plants. He says the drug can block thyroid gland from accumulating radioactive wastes such as resulted from the TMI accident. (HP)

10/1/79 "TMI: A Hot Attraction."
This brief article explains TMI's tourist appeal. A few small souvenier items are described. It concludes with a short explanation of the cleanup problems and future plans to reactivate the reactor. (NW)

10/2/79 "NRC Nixes Discharge Into River."
The NRC sees no possibility of permitting discharge of radioactive wastes into the Susquehanna River. (HP)

"Did Met Ed, NRC Downgrade Data on Core Readings?"
Did the initial data indicate the severity of the core problem? Why did Met Ed and the NRC wait 48 hours to report the data? Was there a cover-up? These questions are being discussed by the Kemeny Commission and the Sen. Subcommittee on N Regulation. (LIJ and NYT)

10/3/79 "NRC Water Decontaminant Decision is Expected Soon."
The article discusses the expected NRC decision on the decontamination of water at TMI. The Epicor process is explained. (HP)

"N-Accident Notification Rules Due."
The NRC will propose that N plant operators warn federal, state, and local governmental officials immediately of minor mishaps that might lead to serious accidents. (HP)

10/4/79 "NRC Shifts 'Blame' on Notice to Met Ed."
The NRC levels several criticisms at Met Ed for its performance during the two days following the TMI crisis. (HP)

"Unit I Contamination Could Breed Lawsuit."
The NRC appears to prefer transferring stored water from Unit II to Unit I rather than releasing it into the Susquehanna. If the NRC orders this and Met Ed resists, the NRC anticipates lawsuits from the utility. (HP)

"N-Foes Score NRC Plan to Limit Free Data Distribution."
The ECNP charges that the NRC proposal to limit distribution of free documents will deprive antinukes of information needed to participate in the NRC legal procedures. (HP)

"If Another TMI Occurs, NRC May Take Over Plants."
The NRC commissioners told the Senate Subcommittee on N Regulation that in the case of another TMI they are prepared to move quickly to take control of the damaged plant. (LIJ)

10/4/79 "Mayor Wants NRC Delay on TMI Water Treatment."
Mayor Wohlsen wrote the NRC that Met Ed is using the leak and possible threat of future contamination to force a hasty approval of the Epicor II treatment system. A letter to the NRC by Arnold of Met Ed urging adoption is also cited. (LIJ)

10/5/79 "NRC Delays TMI Water Cleanup Ruling."
Mayor Wohlsen and Attorney Kohr for the SVA are pleased to learn that the NRC staff has decided to delay authorization for the startup of the decontamination of radioactive water. (LIJ and HP)

"N-Damage Law of U.S. 'Fuzzy.'"
Mayor Robert Reid told the Pa. Commission on TMI that the federal law covering property damage from N accidents is 'fuzzy.' He said there is a need for information on the plant cleanup. The article also indicates that the initial estimates of radiation were high. (HP)

"Weigh Stress From TMI, NRC Told."
The Thornburgh administration says the federal government has legal obligation for psychological stress that activation of TMI may cause area residents. (HP)

10/6/79 "TMI Dumping Near OK in State of Wash."
Washington State officials give permission for tractor trailers loaded with TMI waste to proceed to a dump in Richland. Gov. Dixy Lee Ray had ordered the dump closed earlier. (HP)

10/9/79 "Group Urges Hearings on TMI Flight Hazard."
The ECNP urges the NRC to hold hearings on the hazards caused by aircraft flying over TMI. (HP)

10/9/79 "Rate Hike Bid Slated by Met Ed."
Met Ed cites cash flow problem as the basis for asking the PUC for a 10-12% rate hike. (HP)

10/10/79 "GPU Emphasizes Investor Rights."
If the NRC attempts to enforce a transfer of radioactive waste water from Unit II to Unit I, Met Ed will act to protect the interests of its investors. (HP)

"Members of PANE Slate Meeting Tues."
PANE's participation in NRC intervention hearings in November will be the subject of a local meeting. (MPJ)

"Royalton's Citizens Hear From 2nd NRC Rep."
A NRC spokesman addressed a Royalton Council meeting and only a few citizens are present. (MPJ)

10/11/79 "No Decision Made by NRC on TMI Water."
The NRC met for four hours without reaching a decision on use of a controversial water treatment system at TMI. The Pa. DER testified in favor of Epicor II. The city of Lancaster, SVA, and F&M Geology Professors Arthur H. Barabas and Steven Sylvester opposed the Epicor II system. (LIJ and HP)

Editorial: "No to a Second Chance."
The editorial states that GPU and Met Ed had the chance to operate a N power plant on TMI. He feels they did not do so safely and there is not a single persuasive argument for giving them a second chance. (LIJ)

10/12/79 "NRC Said to Back 'Epicor II' for TMI."
Although not announced, it is believed that the NRC favors the Epicor II system cleanup for TMI. (HP)

10/12/79 "U.S. Panel Finds Problems Persist on Atom Power."
A report by the Rogovin Panel says the NRC's emergency response team remains inadequate six months after the accident at TMI. (NYT)

10/13/79 "Gekas: N-Plan Lacking."
State Sen. George Gekas says there was no statewide emergency plan with NRC approval at the time of the accident. (HP)

"U.S. Judge Dismisses TMI Suit."
US Middle District Court Judge dismissed 'for lack of subject matter jurisdiction' a request for an injunction by the SVA against dumping wastewater into the Susquehanna River. (HP and LIJ)

10/14/79 "Nuclear Aide Sees Gains in Handling Emergencies."
Inadequate emergency response charges are denied by the NRC. (NYT)

10/15/79 "More Nuclear Woes."
Some problems experienced in N plants across the country are discussed. Troubles such as radiation leaks, 'sloppy' waste disposal, faulty packaging and transporting of radioactive materials, and falsified documents affecting plant construction are cited. Regarding TMI, the article explains the confusion and unwillingness of the plant's operators to believe in the accuracy of their gauges, which then led to a series of errors and the accident. The NRC is reportedly expected to undergo structural changes; however, its pressing problem is the disposal of TMI's radioactive water. (NW)

10/16/79 "TMIA Teach-In Set in November."
TMIA and the PIRC will sponsor an anti-N teach-in at HACC. (HP)

10/16/79 "Restarting for TMI NRC Topic."
The NRC will meet in Middletown with representatives from GPU and Met Ed to discuss the restart of Unit I. (HP)

"Swatara Plans Poll on the Future of TMI."
The Swatara Township commissioners agreed to poll the residents on the future of TMI, but refused to act on a resolution to close the plant. (HP)

10/17/79 "NRC Orders Epicor Filtering of TMI Radioactive Water."
In an unanimous order, the NRC commissioners agreed with staff that the use of the Epicor II process "will not have a significant effect on the environment." SVA is asking for a federal judge to issue a temporary restraining order. (LIJ and HP)

"Met Ed Ad."
The ad announces public meetings on TMI Unit I restart hearings. (LIJ)

"Many Haven't Joined the Ranks of the Anti-Nukes."
A report of a Rutgers University study of the residents in the TMI area. The author concluded that the area is less anti-N than one would expect. (MPJ)

10/18/79 "NRC Gets a Storming Reaction on Re-Opening of TMI's Unit 1."
Over 200 citizens registered strong opposition to the proposed return to operation next year of TMI's Unit I. Herman Frimm, a retired physicist from Annville, spoke in favor of the restart. Representatives of the SVA and Mrs. Coder of the LWV of Lancaster County spoke in opposition. (LIJ)

"Epicor II; Awaiting Decision From Judge."
Unless SVA's request for a temporary restraining order is granted, the

Epicor II water treatment system will begin operation immediately. (LIJ and HP)

10/19/79 "Epicor II Turns on Monday."
A three judge panel in Philadelphia refused to grant an injunction to prevent the operation of Epicor II on the grounds it was not "an immediate threat to the environment" since the water will not be discharged into the Susquehanna River after processing. (LIJ and HP)

"Women Gather to Hear Nuclear Power Promoted."
NEW, an organization assisted in its educational efforts by the AIF, sponsored forums around the country on 10/18 as part of N Energy Education Day. (NYT)

10/20/79 "Halt is Urged on N-Reactors."
The Kemeny Commission has voted to recommend a halt in the construction of new N reactors. (HP, LIJ, and NYT)

"Cleanup at TMI to Begin."
TMI plant officials ready plans for the decontamination of radioactive water. (HP)

10/21/79 "3 Mile Island Peril Withheld by Aids at Reactor."
NRC's panel on the TMI accident hears testimony from a plant supervisor, Brian Mehler, indicating that some supervisor's at the plant were aware on the first day that the reactor core was uncovered. They waited until the 2nd or 3rd day to give this information to the NRC. Arnold of Met Ed disputes this testimony. (NYT)

10/22/79 Editorial: "Epicor II."
The editorial reiterates the paper's opposition to the restart of the TMI reactors or any release of the contaminated air or water into the environment as part of the TMI cleanup. The editor

feels that the continual buildup of radioactive water at TMI is a greater threat than the use of Epicor II system. (HP)

10/22/79 "Blame It On TMI."
The article discusses the TMI Syndrome in Harrisburg. TMI seems to have been blamed for bureaucratic problems not connected with the crisis. (HP)

Editorial: "Monitoring Needed."
The editorial rejects the idea of the State BRP that their monitoring system is adequate. The editorial argues for local control and continuous monitoring since neither Pa. nor the federal government appear likely to be of much help. (LIJ)

10/23/79 "3 Mile Island Panel Critical of Utility & U.S. Regulators."
The Kemeny Commission concludes that Met Ed did not have knowledge, expertise, and personnel to operate or maintain TMI adequately. (NYT, HP, and LIJ; also 10/24/79 MPJ)

"Air Lock Forces Epicor Shutdown."
The Epicor II system was shutdown twice by an air lock in an underground line carrying radioactive water from an auxiliary building. (HP and LIJ; also 10/24/79 NYT)

"N-Ban, Governor's Aid Asked."
A coalition of anti N groups asks Gov. Thornburgh to endorse a ban on all aspects of N energy in Pa. The Governor prefers to wait until all studies are complete. (HP)

"N-Energy Filmstrip Provided by Patriot News."
N power and safety is the subject of a contemporary news filmstrip provided by the Patriot News. The controversial

points of the filmstrip and its usage are discussed. (HP)

10/24/79 "Epicor II Idle as Technicians Check System."
Technicians are attempting to solve the problems of Epicor II. (HP and LIJ)

"N-Foes Unahppy With NRC Offer."
ECNP expresses displeasure at NRC's decision to hold hearings on Epicor II after approving its use. (HP)

"Steelton Council Studying Petition to Keep TMI Closed."
The Steelton Borough Council is under pressure from some residents to pass a resolution calling for the permanent closing of TMI. (HP)

"Citizens Respond Heatedly to Unit I Reopening."
Over 300 people crowd a local fire hall for NRC technical hearings on the restart of Unit I. The article includes quotes from citizens. (HP)

"PANE's Lawyer Addresses Group."
A lawyer reports at the Middletown meeting of PANE relative to efforts against the reopening of TMI. (HP)

"TMI Happenings."
A listing of area meetings concerning TMI is provided. (MPJ)

"TMIA Group Bake Sale."
The TMIA sponsors a bake sale for the organization's legal fund. (MPJ)

"NRC Prehearing on Unit I Restart is Scheduled."
A prehearing conference relative to the restart of Unit I will be held in Harrisburg. The key issues are identified. Board members are listed. (MPJ)

10/25/79 "N-Plant 'Never Seriously Weighed.'"
The article reports that the Kemeny Commission never seriously considered the option of phasing out N plants for safety reasons. (HP)

"GPU Units Ask Lower Price on Purchased Power."
The GPU asks the PUC to reduce the cost of electricity that GPU is purchasing from other utilities to replace the power lost as a result of the TMI accident. (HP)

"Utilities Seeking to Keep Unit I in Rate Structure."
Met Ed and the PEC tell the PUC that their customers should continue to pay costs associated with Unit I, even though the unit is not now operating. (HP)

"Bubble Gone, Epicor Fixed."
The Epicor II system is in operation again following the elimination of a bubble that clogged a pipe. (HP and LIJ; also 10/27/79 NYT)

"Carter Panel Would Ban New Reactors in States Lacking Emergency Plans."
The Kemeny Commission will recommend to Pres. Carter that the approval of local and state emergency plans be a precondition for the construction of new N reactors. (LIJ)

"Power Sale to Met Ed Changes to Cost PP&L Customers More."
The PP&L has filed a petition with the PUC asking permission to alter energy charge costs so that its customers would pay instead of Met Ed for the $14 million charge. (LIJ)

10/26/79 "Kemeny Sought N-Plant Moratorium But Lost."
The Kemeny Commission report will not propose a moratorium on the issuing of

N power permits. Chairperson Kemeny had supported this, but lost. He felt it would have shown muscle. (HP)

10/26/79 "'Class 9' Step Dim at Unit 1 Hearings." The NRC will probably oppose general discussion of class 9 accidents on the assumption that chances of such an event are too remote to consider. (HP)

10/27/79 "Epicor Handling of TMI Water Seen Promising." Early analysis of water processed by the Epicor II system indicates that when fully diluted, the water will meet standards for discharge into the Susquehanna River. (HP)

"Met Ed Penalized." The NRC fined Met Ed $155,000 for seven specific violations and holds open the possibility of revoking its license. (HP, LIJ, and NYT)

10/29/79 "8 Ask to Be Heard on Customer-Paid TMI Costs." Eight groups have petitioned the PUC to participate in hearings on whether customers of Met Ed and PEC should pay costs associated with the Unit I reactor. (HP)

10/30/79 "N Evac. Discussed." Londonderry Township officials have developed a proposed plan to be implemented in the event of a mishap at TMI. (HP)

10/31/79 "Report Replies Mixed." The article reports the initial reactions of politicians and N related groups to the Kemeny Commission report. (HP)

"House Fails to Revive TMI Select Committee." The Pa. State House blocked efforts by democrats to extend the life of the

Select Committee on TMI for another year and to give it additional powers. (HP)

10/31/79 "N-Panel Pushes Rules Tightening."
The Kemeny Commission says the accident was not serious or broad enough to justify curtailing the N power industry. It urges that the NRC be restructed, plant operator training be toughened, utility expertise be deepened, and uniform emergency plans be developed. (HP, LIJ, MPJ, and NYT)

"Anne Trunk Can Live With TMI and N-Power."
Anne Trunk has reached the conclusion that N power plants can be safe. However, utilities should be placed on probation for two years, and "if they aren't able to come to a safe level, then their license should be taken away." (LIJ)

"House Ok's Training Program for Nuclear Plant Operators."
The article details a US House program for training federal N power plant operators. The plan is said to be a direct result of the TMI mishap. (MPJ)

"TMI Happenings."
A listing of area meetings concerning TMI is provided. (MPJ)

"PANE Meeting is Set."
Middletown area PANE will meet. A film on N energy is to be shown, followed by a question/answer session. (MPJ)

"Statement Claims Unit 1 Closed Due to Discrimination."
Met Ed charges that Unit I is out of service because it is an object of discrimination by the NRC. The article also discusses the relation of Unit I to Met Ed's rate base. (MPJ)

11/1/79 "Proposal for New NRC Resisted."
The Senate and House subcommittees react well to the Kemeny Commission recommendations. They seem uneasy about the proposal for a strong administrator to replace the NRC five member board. Some are critical of the lack of a moratorium on plant construction. (HP and NYT)

"GPU Objects to Portions of Report."
The GPU reacts critically to the Kemeny Commission's report, but believes that the report will strengthen the N industry in the long run. (HP)

"Met Ed Says Governments Will Be Paid."
Met Ed will reimburse local and county governments within a 10 mile radius of TMI for some extraordinary expenses. (HP; also 11/2/79 LIJ)

Editorial: "Little New and Little to Cheer."
The editorial states that the Kemeny Commission report has little new and little to cheer. It stresses that those on the fence awaiting the report must now act. It says the report did not answer whether a future serious accident is inevitable. (HP)

"Nuclear Report Mixed, Power Industry Finds."
The N power industry expresses relief that the Kemeny Commission did not propose a moratorium on the licensing of N plants. (NYT)

11/2/79 "Utility at 3 Mile Island Ordered to Prove Its Right to Keep License."
The Pa. PUC orders Met Ed to show why its license to sell electrical power should not be revoked. (HP, LIJ, and NYT)

11/2/79 "House Panel Gets Rough With NRC's Hendrie."
A House subcommittee gives a rough time to NRC Commissioner Hendrie. They attempted to get the NRC members to admit to its inadequacy. (HP)

"N-Foes Appeal Cleanup Ruling."
SVA has asked a federal court to overturn the dismissal of the group's attempt to force the NRC to prepare an EIS before allowing further decontamination work at TMI. (HP)

"Infant Death Rise Disputed."
Ernest J. Sternglass said TMI caused the rate of infant deaths to triple at Harrisburg and Holy Spirit Hospitals for the three month period beginning 5/79. Hospital statistics and officials dispute this charge. (HP)

"Met Ed to Pay Expense Money to Local Towns."
Met Ed expects to pay between $100 and $200,000 to reimburse communities for some expenses incurred during the accident. Overtime for policemen is one example of such reimbursement. (LIJ; also 11/7/79 MPJ)

11/3/79 "NRC Executive on Way Out."
The NRC has decided to dismiss its Executive Director for Operations, Lee Gossick. (HP)

"Prehearing Rounds on TMI Scheduled."
The NRC has scheduled prehearing conferences for persons involved in the forthcoming hearings on the proposed reactivation of Unit I. (HP)

11/5/79 "Kemeny Raps N-Industry Attitudes."
Kemeny says that if the N industry does not change its attitudes, public opinion will destroy it. (HP)

11/5/79 "Met Ed, NRC Withheld News."
The Kemeny Commission TSAR on the Public's Right to Information concluded that during the height of the accident, "Met Ed and the NRC withheld news from the public." (LIJ and NYT)

11/6/79 "Panel to Weigh Closing Reactors Near Big Cities."
Chairman Hendrie says the NRC will have to determine whether some of the nation's 72 operating plants might have to be closed because of their proximity to population centers like NYC or Chicago. (NYT and HP)

Editorial: "Sternglass Study TMI Needs No Fabrication."
The editorial criticizes Ernest Sternglass for his analysis of infant deaths in the TMI area. It states that the area does not need any more stress. (HP)

"Derry Denies TMI Appeal."
Derry Township supervisors refused an appeal sponsored by the HAAANP to adopt a resolution calling for the permanent closing of TMI. (HP)

"Royalton Divided on TMI Inquiry."
Royalton residents responded to a questionnaire on TMI. A majority did not evacuate during the crisis; they were evenly divided on whether the plant should be reopened. (HP)

11/7/79 "Unit 2 Switch to Coal Eyed."
The GPU is studying the feasibility of converting Unit II into a coal-fired plant. (HP and LIJ)

"Area Anti-Nuclear Group Seeks Hearings on Epicor."
The SVA asks the NRC to hold hearings on Epicor II system at TMI. (HP)

11/7/79 "TMI Panel Discussion Planned."
Capital Area Council on Social Studies will sponsor a conference titled 'Coping with the TMI Crisis.' (MPJ)

"PANE Meeting to Feature Professor."
The article reports on the agenda of a PANE meeting. (MPJ)

"Public Meeting on TMI Cleanup."
The DER will moderate the third meeting in a series of public briefings on the cleanup of TMI. (MPJ)

"TMI Happenings."
A listing of area meetings concerning TMI is provided. (MPJ)

"TMIA Petitions NRC for Halt to Hearings."
The TMIA filed a motion before the NRC to stay all proceedings for an indefinite period in light of the Pa. PUC decision to hold hearings on Met Ed's ability to operate within the Commonwealth. (MPJ)

"Lobbyist to Speak on Energy Issues."
Drew Diehl, Sierra Club lobbyist on energy issues, will speak on the mood in D.C. since the TMI accident. (MPJ)

"ACRS Schedule Nov. 8-10 Meetings."
The ACRS will hold a technical meeting relative to the application of the TMI experience to other N reactors. (MPJ)

"GPU Chairman Responds to Findings of President's Investigative Commission."
GPU responds to the Kemeny Commission Report by saying its conclusions support the belief that the accident involved the entire industrial, technological, and regulatory structure of N power. (MPJ)

11/8/79 Editorial: "PUC Failed to Probe Hazards."
The editorial praises the PUC for raising to the challenge of TMI. However, it voices disapproval for the

agency's approval of a high voltage transmission line. The editor feels that the high voltage lines pose various hazards that must be probed. It states that the lessons of TMI include being complacent in accepting assurances of safety in technology and dismissing trouble warnings too soon. (HP)

11/9/79 "Weigh Anguish, N-Panel Urged."
The NRC ASLB was told by a citizen group that it should consider psychological stress factors in the decision of reopening Unit I. (HP)

"PUC to Hear Met Ed Case."
The PUC will begin TMI related hearings. Issues include: 1) Met Ed's operating license, 2) a surcharge rate hike request, and 3) whether Unit I should be taken out of the rate base. (HP)

"TMI Doubts Affect Cleanup."
The loss of public confidence is causing problems for utility and governmental officials in making the cleanup decision on Unit II. (HP)

"3 Mile Island Executive Urges U.S. Role in Cleanup."
GPU's president says the company is seeking help from DOE and the Electric Power Research Institute to reduce its cost for cleanup. (NYT and LIJ)

"Mayor Tells Capitol Panel: NRC, TMI Still Resistive."
In testimony before the Senate Subcommittee on N Regulation, Mayor Wohlsen testified that "both Met Ed and the NRC are still resistive to good, tough public review of the cleanup operation." (LIJ)

11/10/79 "NRC Chief Against TMI Federal Aid."
In testimony before a Senate Environmental Subcommittee, Hendrie estimates that it will take as long as five years to clean up the crippled TMI N power

plant. He is opposed to federal aid. Aaron Levy, Director of Corporate Regulation for the SEC, states that TMI can afford to pay the estimated $800 million in cleanup costs. Dieckamp states that Met Ed is seeking a large DOE research grant. (LIJ)

11/10/79 "Consolidation Plan Rejected by Anti-Nukes."
Attorneys of Met Ed have suggested that the anti N groups consolidate their presentations to the NRC. The groups reject the attorneys' proposal. (HP)

11/13/79 "N-Hearing is Slated."
The NRC will hold a one day informal hearing in Harrisburg for public statements on whether TMI should be called an 'extraordinary N occurrence.' (HP and LIJ)

"TV Gets Radiation Readings."
A small TV camera and a strobe light were inserted inside Unit II to videotape radiation readings. Gamma levels were down about a thousand times from the level immediately following the accident. (HP, LIJ, and NYT)

"University Women Back City on TMI."
The Lancaster Chapter of the AAUW, voted unaminously to support the city's lawsuit. (LIJ)

11/14/79 "SVA Argues in Court to Reinstate Lawsuit."
Lawyers for the SVA have argued before a three judge panel of the US Third Circuit Court of Appeals in Philadelphia to have their lawsuit reinstated. (LIJ)

"Survey on TMI Plant is Pondered."
The Mechanicsburg Council is considering undertaking a survey before it takes a stand on the reopening of TMI. (HP)

11/14/79 "NRC Balks at Suggestion of Lesser Role." The NRC is in agreement with most of the recommendations of the Kemeny Commission, but rejects those suggesting it be stripped of authority in future emergencies. (HP and NYT)

"Met Ed Hearing Date Set." The PUC sets 12/10 as the date for the start of hearings on the future of Met Ed and TMI related issues. (HP and MPJ)

11/15/79 "Unit 2 Venting Asked." Met Ed asks the NRC for permission to vent Krypton 85 into the atmosphere. (HP)

"PUC Control Over Power Plant Construction Supported." The PUC now finds out about new power plants after construction has begun. A bill giving the PUC control over construction is endorsed by several groups. (HP)

"TMI Owners Hire Kline as Consultant." The GPU has retained the consulting firm of former Lt. Gov. Kline to counsel them on TMI PR matters. (HP)

"NRC Staff Will Conduct EIS." The NRC will conduct a study on the implications of the decontamination and recovery operations at TMI. (HP)

"Carter, Denouncing Terror, Warns Iran on Hostages." Pres. Carter endorses N energy as part of his comprehensive energy program in a speech to the AFL-CIO convention. (NYT)

"Met Ed Wants to Vent Radioactive Gases into Air." Fearing a leak or an accident in the storage facilities, Met Ed wants to vent radioactive gases from TMI next year. (LIJ; also 11/18/79 NYT)

11/16/79 "Citizens Give Their Views on Unit I Reactor's Future."
An account of public presentations to the NRC administrative law panel on issues surrounding the future of Unit I are provided. (HP)

"Ertel Dubious on N-Report."
Cong. Ertel states that the Kemeny Commission report is inconsistent and does not add to N safety. (HP)

"3 Mile Island: A License at Issue."
Bankers, analysts, and utility executives are expressing concern over the forthcoming Pa. PUC hearings into whether Met Ed should retain its license. (NYT; also 11/17/79 LIJ)

11/17/79 "NRC Hears the Cries of 'Abandon TMI.'
Anti-TMI groups dominate a hearing of the ASLB in Hershey. (LIJ and HP)

"State Warned on TMI Radioactive Gas Risk."
In an informal briefing with Gov. Thornburgh, Met Ed officials warn the state about the need to vent radioactive gases. (LIJ and HP)

11/19/79 "Hundreds Rally in Baltimore Against Atomic Waste Plans."
Hundreds of people rally in Baltimore to protest a proposal by Met Ed to dump decontaminated water from TMI into the Susquehanna River. (NYT)

11/20/79 "TMI Malfunctions as Election Issue."
The TMI accident did not appear to be a factor in area elections. (HP)

Editorial: "Radiation Fraud."
The editorial raises questions about Met Ed's claims that Krypton release will result in low radiation exposure. A German study challenges the basis NRC uses to estimate dosage. The study may

function to question the credibility of NRC estimates. (HP)

11/20/79 "Venting of TMI Radioactive Gas May Hinge on Environmental Study."
R.H. Vollmer, NRC's Staff Director for TMI Operations, indicated that an EIS would first be undertaken before deciding the venting of radioactive gases. (LIJ)

11/21/79 "Group Will Oppose TMI Conversion."
Air pollution is cited as a basis for opposition of the Swatara Water Sports Association to converting TMI from a N to a coal power plant. (MPJ)

"GPU Claims TMI Clean Up is in Residents Best Interest."
Utility officials meet with Gov. Thornburgh to discuss the TMI cleanup and decontamination. They restate the priority of the cleanup and the commitment of resources to it. (MPJ)

"TMI Happenings."
A listing of area meetings focusing on TMI is provided. (MPJ)

"Extension of Time Given to Met Ed."
The NRC has granted Met Ed until 12/4 to respond to notice of violations and proposed imposition of civil penalties in connection with the TMI accident. (MPJ)

"TMI Transcripts are Available."
Copies of telephone conversations recorded during the TMI crisis are available at the NRC regional offices. (MPJ)

"Welfare Agency Plans Study on TMI Stress."
Pa. DPW will begin a study on the mental health impact of TMI over a 10 mile radius. (HP; also 11/28/79 MPJ)

11/22/79 "NRC Unit Hears Testimony."
An informal meeting is held in Harrisburg by the NRC on whether the TMI accident should be termed an 'extraordinary N occurrence.' (HP)

"Met Ed Set to Carry Out Panel's Recommendations."
Met Ed assures the PUC that it will carry out all recommendations of the Kemeny Commission that relate to the safe operation of TMI. (HP; also 11/28/79 MPJ)

11/23/79 "Three Mile Island to Get Environment Statement."
The NRC is to propose an EIS on the decontamination of the damaged N reactor at TMI. (NYT; also 11/28/79 MPJ)

"Md. Mulls Dumping Lawsuit."
Md. is considering going to court to stop the dumping of contaminated TMI waste into the Susquehanna River. (HP and LIJ)

11/24/79 "NRC Chief Will Head Up Emergency Unit."
Joseph Hendrie plans to make himself head of the NRC emergency response team. The move is viewed as an attempt to meet criticism of the Kemeny Commission. (HP)

11/27/79 "TMI Role in Middletown's Future 'Small.'"
A consultant states that TMI will have little effect on the future growth and development of Middletown Borough. (HP)

"Township Rejects Krypton Release."
Newberry Township supervisors approve a resolution opposing Met Ed's proposal to release Krypton into the atmosphere. (HP)

"Survey Planned at Mechanicsburg."
The Mechanicsburg Borough Council will meet to decide what questions will appear in a survey on TMI to be sent to all Borough residents. (HP)

11/28/79 "Weigh Rate Hike First, Met Ed Asks PUC." The PUC will hold hearings on Met Ed's right to continue operations. The question of an interim rate hike will be considered first. (HP)

"Resident Asks Council to Oppose Krypton Venting."
A Middletown resident, Donald Hosaler, requests that Borough Council pass a resolution opposing the venting of Krypton from Unit II. (MPJ)

"Boro Reimbursed: Fire Companies May Get Donation."
The article details reimbursement by various agencies of $16,860 for facilities used and services rendered during the TMI crisis. (MPJ)

"TMI Happenings."
A listing of area group meetings relative to TMI is provided. (MPJ)

11/29/79 "3 Mile Island Utility Shows Drop in Net."
Met Ed reports net income of $86.7 million for the 1st 10 months. This represents a 23.1% decline. (NYT)

"NRC Unit Sets Epicor Session."
At the request of SVA, the NRC will hold a prehearing in Harrisburg on the use of Epicor II system. (HP)

11/30/79 "TMI Ships Out Nuclear Waste."
TMI waste is trucked to Hanford, Washington. (LIJ and HP)

"Rap at TMI Operators Queried."
An independent NRC advisory group criticizes previous NRC report that put the blame for the TMI crisis squarely on operators. The group feels other factors were not considered fully. (LIJ)

11/30/79 "Freeze on Building of Nuclear Plants Rejected by House."
The House rejects a six month moratorium on N reactor construction. (NYT and HP)

12/2/79 "Nuclear Safety Campaign Aims at the Old Reactors."
The article reviews NRC plans to upgrade safety of existing N power plants. The NRC and the UCS differ on the value of the plan. (NYT)

12/4/79 "City Wants Hearing Role on TMI Plan."
Lancaster City has asked the PUC for the right to participate in their hearings dealing with the future of Met Ed as a licensed utility in Pa. (LIJ)

12/5/79 "Public Hearings on TMI Clean Up Slated December 11th."
The Pa. DER will moderate the fourth public hearing on TMI cleanup in Harrisburg. (LIJ and MPJ)

Met Ed Ad: "Progress Report-The Clean Up."
An ad extolling the progress Met Ed is making in the TMI cleanup. (LIJ)

12/6/79 "Emergency Plans Approved by Federal Nuclear Agency."
The NRC tentatively adopts new rules that would require N power plants to have federally approved state and local emergency plans or face closing. (NYT and LIJ)

"Met Ed Asks for Review of NRC Fine."
Met Ed argues that deficiencies were due to the lack of knowledge and not to deliberate acts. Therefore, they are asking for their fines to be reviewed. (LIJ and NYT; also 12/12/79 MPJ)

12/7/79 "Videotape Film Shows No Structural Damage to 3 Mile Island Plant."
Met Ed says a videotape of the crippled reactor indicates no structural damage. (NYT)

12/8/79 "President Carter Dismisses Hendrie as Chairman of NRC."
Pres. Carter plans to reorganize the NRC and still indicates a need for N energy. (NYT)

"Met Ed Pegs TMI Gas at Lower Level."
Met Ed officials estimate that the level of radioactive gas inside reactor II may be less than previously estimated. (LIJ)

12/10/79 "SVA Hires Consultant."
The SVA has retained Gregory C. Minor of San Jose, California to help the group prepare evidence and testimony concerning NRC environmental impact hearings about cleanup. Beverly Hess of SVA cites a University of Heidelberg study indicating that the amount of radiation released due to the accident at TMI was higher than official estimates. (LIJ)

"TMI Has Turned Quiet Types to a New Activism."
A feature story about two anti N families who are members of ANGRY. (LIJ)

12/11/79 "Truck Carry TMI Waste to Dump Crashes."
A truck accident occurred in Missoula, Montana but the containers are still intact and there was no radiation danger to the public. (LIJ and NYT)

"Weekend Hours at TMI."
Observation and exhibition center at TMI is opened during the weekend. (LIJ; also 12/12/79 MPJ)

"Met Ed's Future at Stake as PUC Launches Hearings."
Testimony pro and con is presented at the hearing. (LIJ)

12/12/79 "Power-Short Met Ed Lists 1980 Needs."
Met Ed needs 2.7 billion KW of outside power to meet demands caused by the TMI units being out of service. (HP)

12/12/79 "Connally, in Altoona, Says N-Power Boost is Needed."
John Connally, a republican presidential candidate, says he would cut 'red tape' and build more N power plants. (HP)

"Keeping TMI Unit 2 Idle Unpatriotic."
B. Cherry, V.P. of GPU, testified before the PUC that the starting of the undamaged reactor would save $168 million in 1980 in the purchase of foreign oil. Therefore, the undamaged reactor should be permitted to restart. (LIJ)

12/13/79 "Officials Say TMI-1 Hearings 'Unnecessary.'"
Arnold states that the NRC hearings on the safety of the undamaged reactor at TMI are 'unnecessary and inappropriate.' (LIJ and HP)

"Met Ed Sees Gas Release in 1981 to Aid Clean Up."
In public hearings in Maytown, Met Ed officials state that more radioactive gases may have to be released when technicians remove the uranium fuel core from the damaged reactor. (LIJ)

"Questions Remain on TMI Radiation Dose."
In testimony to the PUC, Patricia Longnecker of Elizabethtown, indicated that she possessed evidence that radiation releases during the height of the accident were far greater amounts than the government had reported. (LIJ)

"U.S. Panel Gets Atom Plant Plan."
The staff of the NRC outlines a 5 year plan to improve the safety of N power plants. It is expected that the plan will cost the NRC $170 million and the industry $1.6 billion. (NYT)

12/14/79 "In TMI Shadow, Brown Demands No More Nukes."
Gov. J. Brown, in front of TMI, assailed Pres. Carter as the 'chief salesman for

the N industry' and called for a halt to the further development of N power because it threatens 'the genetic future of the human species.' (LIJ)

12/19/79 "Rate Hike Case Counsel Warned."
A PUC official warns opponents of the Met Ed rate hike to prepare a more meaningful case. (HP)

"York Firm Tells PUC Met Ed is Reliable."
At the PUC hearings to determine whether Met Ed should stay in business, the York industrial community gave Met Ed a vote of confidence by saying it is a reliable firm. (LIJ)

"If Not Safe, TMI Will Remain Closed."
Dr. Frank Press, Chief Science Advisor to Pres. Carter, said that if TMI is not safe, it will not be reopened. Discussion of Middletown meeting on TMI reported. (MPJ)

"Decontamination Schedule Released by Met Ed."
The highlights of Met Ed's plan to decontaminate and defuel Unit II are discussed. (MPJ)

"NRC Announces Changes on Emergency Planning at Nukes."
As a result of TMI, the NRC has proposed new regulations to upgrade requirements for emergency planning in the area near N power plants. (MPJ)

"7.5 Million Upgrade for TMI's Fire Protection System."
Improvements in TMI fire protection are reported and other plans are projected. $7.5 million will be spent on Unit I, of which two thirds is for fire protection. $2.5 million will be spent on Unit II. (MPJ)

12/20/79 "TMI Faces Suit Over Lost Baby."
Ed and Perri Klick of York County plan to file a suit in the US District Court in Harrisburg. The suit will be filed against Met Ed, GPU, and the plant designer B&W. The suit will claim damages for a stillborn baby. (LIJ; also 2/28/79 HP)

"Does TMI Have Too Many Watchdogs?"
The US EPA and the Pa. DER are monitoring TMI. (LIJ)

"State Nuclear Critics to Join in NRC Rules."
The ECNP will join the NRC proceedings to examine if radioactive wastes can be disposed of safely and permanently. (HP)

12/22/79 "PUC Rebuffs Met Ed Plea."
The PUC says it will not consider Met Ed's request for a $55 million rate hike without at the same time considering whether the operating license should be lifted and what effect the inoperative Unit I should have on rates. (HP)

"Panel Asks Better TMI Safety System."
A special NRC Panel studying radiation safety released a report calling for better safety systems as a result of the TMI accident. (HP)

1/2/80 "NRC Panel Says Met Ed Radiation Safety Flawed."
The NRC states that Met Ed's program for cleanup of TMI has deficiencies and needs upgrading to assure the safety of its workers. (HP and MPJ)

1/3/80 "Nuclear Plants Drag on Compliance."
The article concludes that short term requirements for N safety in the wake of TMI have been met in only half of the nation's power plants. (HP)

1/3/80 "Md. Eyes Legal Step on TMI."
Md. tells the NRC it may go to court to have its TMI cleanup concerns addressed. (HP and LIJ)

"Walker: Keep TMI Nuke, Give Met Ed Boot." Cong. Walker states that he has no objection to the reopening of TMI as a nuclear powered plant. He believes that Met Ed should lose its license because, "Met Ed still hasn't shown me that they are willing to assume the public responsibility that I think should be required of a N operation." (LIJ)

1/4/80 "No Desire to Take on TMI, States PP&L." Because of a PUC hearing on the future of Met Ed as a licensee, PP&L issued a statement indicating that it is not interested in taking over Met Ed's service area. (LIJ)

"Met Ed Says $207 Million Needed to Bring Back TMI."
A PUC staff attorney Joseph Malatesta estimates that $207 million are needed to restore Met Ed to its preaccident financial status. (LIJ and HP)

"TMI Gets Recovery Chief."
Gale K. Hovey, Plant Manager at Allied-General Nuclear Services, is named to head the recovery and restart operation at TMI's damaged Unit II. (HP)

"Intervenors Set in Unit I Case."
The NRC approves a list of intervenors for hearings on the restart of Unit I. (HP)

1/5/80 "'Extraordinary' Finding in TMI Mishap Withheld."
HP learns that the NRC is set to recommend whether the TMI accident is an 'extraordinary N occurence' in the legal sense. (HP)

1/5/80 "City Settles Suit Over TMI Water."
The suit is settled because Met Ed agrees to install and pay for radiation monitoring equipment at Lancaster City's water plant on the Susquehanna River. Furthermore, the NRC also agrees to ban the dumping of radioactive water into the river until 1982, or until an EIS is completed. SVA continues its suit. (LIJ)

1/6/80 "After Three Mile Island."
An article on the future of the N power industry in the US in wake of the accident at TMI is presented in the National Economic Survey. (NYT)

1/7/80 Editorial: "The TMI Settlement."
The editorial criticizes the settlement which led Lancaster to drop its suit since the basic issue of safety has not been addressed. (LIJ)

"City Ponders How To Use Radiation Detection Program."
How the city will utilize the radiation detection program being purchased by Met Ed is still undecided. (LIJ)

1/9/80 "SVA Continues TMI Water Dumping Battle."
SVA decides to continue its legal suit. (LIJ)

"Harold Denton Will Address Area Association."
Denton will speak to MAA on 2/9/80. (MPJ)

"NRC ACRS to Meet."
The NRC's ACRS will hold a technical meeting from 1/10 to 1/12 in D.C. (MPJ)

"Workshops Will Discuss NRC Emergency Planning."
The NRC staff has scheduled four regional workshops with state and local officials and utility representatives to discuss the Commission's proposed requirements to upgrade emergency planning in areas near N power plants. (MPJ)

1/9/80 "TMI-Aircraft Hearing Planned."
The NRC plans to resume hearing on the potential hazard of aircraft flying near TMI. It will also decide if mining and milling of uranium fuel creates a health risk sufficient to prompt revocation of Unit II operating license. (HP)

1/10/80 "GPU Claims Moth Balling of TMI Could be a Very Costly Mistake."
In testimony before the PUC, GPU officials stated that "decommissioning will require the same degree of cleanup that we have undertaken to restore the plant to service." Met Ed estimates $275 million for cleanup, $80 million for purchase of a new fuel core, and $75 million to rebuild the N plant. Therefore, GPU would not save money and might actually spend more if it abandoned the N reactor at TMI. (LIJ)

"Nuclear Peril Minimized."
At the Robert Abbe lectures of the College of Physicians in Philadelphia, Dr. R. Linneman said that the "likelihood of danger from a N accident is minimal." Dr. D. Abramson said that "human nature makes the use of fission too dangerous."

"Nuclear Industry Wants Resumption of Licensing."
The AIF urges the NRC to resume licensing of atomic power plants charging the NRC with engaging in a 'series of endless safety studies.' (NYT)

"Met Ed's Cost Great, Even for Shutdown."
The article discusses the costs involved in the cleanup of TMI whether the plant starts up again or not. (MPJ)

"Met Ed Ability Question Coming."
The NRC agrees to spell out questions regarding Met Ed's technical capacity. These questions must be answered prior to the restart of Unit I. (MPJ)

1/15/80 "Met Ed Goes to PUC With a Plea for Life." Met Ed asks the PUC to terminate its hearings on whether they should be allowed to continue as a licensee. (LIJ and HP)

"NRC Acts to Prevent Radiation Release." The NRC set forth new regulations to prevent release of radiation during a meltdown. This reverses an earlier position that a meltdown was too remote. Following TMI, the NRC staff concluded that chances of a meltdown are higher than previously expected. (HP)

"TMI Exit Payments Top $1.3 Million." The article describes the claims of $1.3 million that have been paid to evacuees who left homes at the time of the TMI crisis. It is the largest single payout in the history of N insurance. (HP)

1/16/80 "NRC to Open Middletown Office." The NRC will open an office in Middletown by the end of the month. (MPJ, LIJ, and HP)

"PANE Slate Jan. Meeting." A PANE meeting is scheduled for 1/23. (MPJ)

"Added Protection Called for in Shipment of Radioactives." New requirement will protect against theft or sabotage of radioactive materials during shipment. (MPJ)

"NRC Panel Makes Recommendations to Commission on TMI Accident." Senior NRC staff recommend that the events on 3/28/79 and thereafter not be classified as an 'extraordinary N occurrence.' (MPJ)

"Public's Aid Sought for Nominee to NRC Board." Nominations to the ASLB Panel are being sought. (MPJ)

1/18/80 "GPU to Create New Firm for Nuclear Facilities."
In testimony before the PUC, Kuhns indicated that GPU will create a new company to operate its N facilities in Pa. (LIJ; also 1/19/80 NYT, and 1/23/80 MPJ)

1/19/80 "State Short on Readiness, TMI Unit Says."
A Pa. House Select Committee on TMI said that Pa. lacks adequate banking laws to aid evacuees, fell short in CD procedures, and was deficient in care of the farm animals. (HP)

1/22/80 "GPU Earmarks $14 Million Toward Unit 2 Replacement."
The GPU has earmarked $14 million toward replacement of Unit II N core. $103 million is budgeted this year for TMI restoration costs. A unit would cost approximately $75 million to replace. (HP)

"NRC Asked to Consider Bubble at TMI an Issue at Hearings."
The NRC was asked to decide if the gas bubble in Unit II is an issue to be considered in the reactivation of Unit I. (HP)

"TMI Owners to Spend $13 Million for Uranium."
The GPU will honor its contract with Anaconda Company to purchase uranium at an average price of $11 per pound. The purchase of the uranium is to encourage banks that GPU is serious in trying to restart TMI and it also could be sold for a profit at a later time. (LIJ)

"TMI Damage Loan Deadline is Jan. 28."
James L. Higgins, the disaster coordinator for the SBA, stated that 1/28 is the deadline for submitting applications for loans for business owners who suffered economic injury as a result of the TMI accident. (LIJ)

1/23/80 "NRC's Advisory Committee Selects New Chairman."
Dr. M. Plesset, Professor of Engineering Science emeritus at California Institute of Technology, was selected as Chairman of the ACRS. (MPJ)

"NRC Will Discuss TMI Decontamination."
The NRC will meet on 1/29/80 to discuss decontamination, EIS, and disposal of radioactive wastes in Harrisburg. (MPJ)

"Regulations Would Restrict Nuclear Material Shipments."
New regulations on the shipment of N materials would limit one shipment to occur for one storage facility at the same time. (MPJ)

1/23/80 "Met Ed Would Prefer 'Mothballing' Option."
A Met Ed spokesman said that if ordered a preferred method for decommissioning Unit I would be to mothball the unit for 30 years and then dismantle it. (HP)

"Lawyer Seeking Wider TMI Probe."
Attorney J. Bowers, representing a group of York area farmers opposed to N power, wants the PUC to subpoena Victor Stello, Director of the NRC's Office of Inspection and Enforcement. The PUC denied his request saying Stello's testimony regarding the causes of the TMI accident can't help the PUC to decide whether Met Ed's license should be revoked. Furthermore, the PUC does not have the authority to subpoena a non-Pa. resident. (LIJ)

"Volunteer Squad to Probe TMI's Containment Building."
Six volunteers are training to become the first humans to enter the containment building at TMI. (LIJ)

"NRC Imposes $155,000 Fine on TMI Utility."
The NRC imposes a $155,000 fine on TMI facility. (LIJ)

1/25/80 "TMI Accident 'Only 60 Minutes From A Meltdown.'"
The primary conclusion of the Rogovin Panel is that the TMI accident came within 30 to 60 minutes of a meltdown. (NYT, HP, and LIJ)

1/26/80 "Federal Magistrate Advances TMI Suit."
A US Magistrate recommends that the court consider a claim for medical detection service for people living near TMI. (HP)

1/28/80 "State, County Differ on Evacuation Plan."
Pa. and Lancaster County officials disagree about the capability of PEMA to conduct an evacuation safely and adequately at the time of the TMI accident. (LIJ)

1/29/80 "Start Cleanup of TMI, DER Head Urges."
Pa. DER urges the cleanup of TMI as the radioactive elements within Unit II are viewed as a long term threat. (HP)

"Energy: A Shift Toward Buying for Oil Reserve."
Pres. Carter's budget for fiscal 81 envisions spending $476 million on projects to make N plants safer. This is a 55% increase. (NYT)

1/30/80 "Utility Ratings Lowered."
Standard and Poor Corporation lowers the ratings of bonds issued by Met Ed. (NYT and HP)

"Perils Seen in Possible Met Ed Reshuffle."
A bankruptcy law expert testified that if the PUC orders a reorganization of Met Ed, the TMI cleanup and decontamination plans could be jeopardized. (HP)

"Met Ed Probers to Enter Unit 2 in Mid-March."
Met Ed intends to send an inspection team into Unit II in mid-March. It is also

reported that the Epicor II is operating less efficiently than planned. (HP)

1/30/80 "Middletown NRC Office Opens Today."
A NRC field office opens in Middletown with John Collins, Deputy Director of the NRC TMI Technical Support Staff, in charge. (MPJ)

1/31/80 "TMI Caused $100 Million Loss in Area."
The accident resulted in a $100 million loss and harmed real estate. Support for this finding is presented in a report to the Kemeny Commission. (LIJ)

"Utility Head Defends Managers of TMI."
In testimony before the PUC, Wm. Lee, Pres. of Duke Power Co. in NC, stated that the "managers of TMI are highly competent and should not be held responsible for the accident at the power plant." (LIJ)

2/2/80 "Few Fears of TMI Down River."
A group of 100 people attended a meeting in Havre de Grace, Md. regarding TMI activities. Minimal fears are expressed but concern did surface over the dumping of the water into the Susquehanna River. (LIJ)

"Met Ed to Prosecute Reporter Who Secured Job as TMI Guard."
Met Ed will prosecute Robert Kapler, a reporter for *The Guide*, a Harrisburg area weekly newspaper, for falsification of documents. He worked at TMI from 1/2 to 1/19 as an undercover reporter to check into the security arrangements at TMI. (LIJ; also 2/4/80 NYT)

2/5/80 "Judge Rules Newspaper Can Print TMI Expos."
A judge of the Dauphin County Court ruled that *The Guide* could print its TMI expose. Kapler wrote "TMI is a paradise for a saboteur, not from without, but from internal subversion." (LIJ and NYT)

2/6/80 "Middletown's Emergency Preparedness Plan Sent to County."
An 83 page document concerning emergency planning, whose completion according to Mayor Reid was, in large part, due to the TMI accident was sent to Dauphin County officials. (MPJ)

"Agreement Reached on Planning for Emergencies at Nukes."
The NRC and FEMA signed a joint statement outlining agencies responsibilities in preparing for emergencies at commercial N power plants. (MPJ)

"Restart of Unit I will be Discussed at Meeting."
A technical meeting of the ACRS will be held 2/7-9 in D.C. on the restart question. (MPJ)

2/8/80 "Payout Omitted by TMI Parent."
GPU voted to omit its quarterly dividend on common stock. (HP, LIJ, and NYT)

"NRC Delays OK of 45 Point Safety Plan."
The NRC has suspended construction and operation licenses for N plants since the TMI crisis. A 45 point safety plan that incorporates suggestions from the Kemeny Commission has been offered, but not approved. Thus, a license freeze continues. (HP)

2/9/80 "PUC OKs Boost in Met Ed Rates."
The PUC granted Met Ed a conditional $55 million rate increase. The hike is intended to help Met Ed purchase energy lost as a result of the TMI accident. (HP)

2/12/80 "Met Ed Faulted on Pacts."
The PUC was told that Met Ed should be penalized for not keeping contracts with coal suppliers that would have resulted in lower prices. (HP)

2/12/80 "City is Cautious with TMI Litter."
Harrisburg's incinerator/steam generating plant has been turning back TMI deliveries except for ordinary litter and refuse. (HP)

"NRC Gets an Earful on TMI."
The article cites comments made by the public and the NRC representatives at a Goldsboro Borough Council meeting. The NRC representatives said that Pa. should teach those concerned with radiation how to read a radiation monitor. (HP)

"TMI Senimar Set for Anniversary of N-Accident."
A day long seminar on lessons learned from TMI will be held at PSU, Capitol Campus, on the first anniversary of the accident. (HP)

"N-Risk Seen as Moral Dilemma."
Jack Holl, designated by US DOE as Project Director for a TMI History, says the moral dilemma of N risk is unsolvable. (HP)

"Post-TMI Changes are Rushed."
The ACRS warned the NRC that the government may be moving too quickly in deciding on the improvements in the wake of the TMI crisis. (HP and NYT)

"TMI Spews Gas Trace at 'Hot' Water Leaks."
Some radioactive gas escaped from TMI on Monday when 1,000 gallons of primary coolant water seeped from a faulty pump into a sump basement of the Unit II reactor auxiliary building. An onsite emergency was declared for almost 6 hours. Company and public officials state "that there is nothing for the public to be worried about." (LIJ and HP; also 2/13/80 MPJ)

2/13/80 "Sen. Heinz Says TMI Covered Up Small Gas Leak."
Sen. Heinz accused the TMI authorities of a cover-up in delaying reports of a small radioactive gas leak at the plant. A report was made public by Heinz who happened to be inspecting the plant that day. (LIJ)

"TMI Neighbors See More Loss of Credibility."
At a NRC hearing concerning an EIS, 125 residents of Middletown voiced fear and confusion as to why Met Ed delayed reporting a small gas leak. (LIJ and MPJ)

"Thornburgh Critical on Leak."
Gov. Thornburgh criticizes Met Ed and the NRC for not promptly notifying him of the leaking primary coolant system at TMI yesterday. (NYT)

"Bechtel to Repair Atom Plant."
GPU names BPC as the prime contractor for cleanup between $400 and $500 million. (NYT)

Editorial: "TMI Rerun: What Haven't We Learned?"
The editorial criticizes Met Ed and the NRC over the failure to communicate honestly regarding the radiation leakage. (MPJ and HP; also 2/14/80 LIJ)

"County Assails Met Ed."
Dauphin County officials complain to Met Ed about the lack of notification during the most recent TMI spill of radiation. (HP)

2/14/80 "TMI Leaks Trace of Gas for 2nd Day."
A gas leak 10 times larger than yesterdays was discovered at TMI. The NRC and the Pa. BRP reported that the three-curie release was insignificant. (LIJ, HP, and NYT)

2/15/80 "NRC Lays Fault for Leaks on Bungling TMI Workers."
Victor Stello of the NRC states that TMI technicians mishandling of a routine gas sampling procedure caused the radioactive gas leak. Stello hedged as to whether Met Ed should be fined. (LIJ and HP)

"NRC Order Affects Met Ed's TMI-2 Operating License."
The NRC has issued an order that would amend the Unit II license. The Unit must be maintained in a safe, stable long term cooling condition. (HP; also 2/20/80 MPJ)

2/16/80 "N-Critics Unaware of Vigil."
Anti-N groups and local politicos state that they have a lack of confidence in NRC monitoring of TMI radiation. They ask Pres. Carter to set up a federal monitoring program. The groups are apparently not aware of Carter's long term environmental surveillance plan for the plant under US EPA. (HP)

"Met Ed Bristles at NRC Job."
Met Ed feels that the NRC has criticized them unfairly for recent events at TMI. (HP)

"Demoted Hendrie Snaps at TMI Safety Setup."
After being demoted by Pres. Carter as head of the NRC, Joseph Hendrie criticized the slowness of the cleanup of radioactive materials at TMI. He feels too much concern is being expressed over the minor releases of radiation. (HP)

"Owners of 3 Mile Island Pays $155,000 Fine for Violations."
Met Ed pays $155,000 fine to the NRC for the violations that led to the 3/79 accident at TMI. (NYT)

2/16/80 "NRC Seeks Means to Speed Clean Up at TMI."
Federal regulators are looking for ways to speed up the decontamination of the crippled TMI N reactor. They indicate some concern that the plant in its present state may pose a safety risk. (LIJ; also 2/17/80 NYT)

"TMI Director Pledges Better Alert System."
Arnold states that "any unusual events at TMI" in the future will be immediately reported to local and state officials. (LIJ)

2/18/80 "NRC Orders New Inquiry About TMI."
Rogovin has been asked to investigate whether Met Ed deliberately withheld information about the potential seriousness of last March's accident. Met Ed repeatedly denies this charge. (LIJ and NYT)

2/19/80 "91 Walk Off TMI Job Over Safety System."
Ninety one electricians walked off the job at TMI in a dispute over radiation safety procedures. (LIJ)

2/20/80 "Contractor Seeks Union Help in Walkout at TMI."
Efforts to resolve wildcat walkout are being initiated. (LIJ)

"Legal Experts Claim TMI Customers Should Pay Most of Accident Bill."
In a hearing before the PUC, George Avery, a Washington lawyer, and Guido Calabresi, a Yale University law professor, argued that the utility customers should shoulder N risk just as they have shouldered other non-N risks. (LIJ)

"Shutdown of TMI is Voted."
The Mechanicsburg Borough Council passed a resolution calling for the shutdown of TMI. (HP)

2/20/80 "And the Leaks Go On."
Radioactive Krypton continues to leak and Arnold states that concern is unnecessary as radioactive gas leaking is normal. (MPJ)

"PANE Meeting Set for Feb."
The PANE meeting for February is announced. (MPJ)

"Rise in Thyroid Problems Suspected Near Atom Plant."
Dr. G. Tokuhata, PDH, reports an apparent increase in hypothyroidism among newborns in the vicinity of TMI. (NYT)

2/22/80 "TMI, Illness Link Denied."
PDH officials react to news reports of cases of infant hypothyroidism. They feel there is no relationship to TMI. Difficulty in establishing norms for incidence of the problem is cited. (HP and LIJ)

"Workers Back on Job at TMI."
Wildcut walkout of workers ends. (LIJ)

"City Agrees to OK Water Monitoring."
Lancaster water authorities took steps to install a radiation monitoring system at the city's Columbia water pumping station. (LIJ)

2/23/80 "Protestors at Mansion."
TMIA and other groups protested for the permanent shutdown of the TMI plant in front of the Governor's mansion. (HP)

"Met Ed Pays Dauphin County for TMI Expenses."
Dauphin County has been paid $22,781.42 for extra costs incurred due to the TMI incident. The county also announces that 20 mile radius evacuation plans are nearing completion. (HP)

2/25/80 "Nuclear Utilities Form Own Insurance Co."
N utilities are forming a new insurance company to make sure they can survive the

financial blow from future power plant shutdowns. Hopefully, the company will be in operation by late spring and will offer maximum coverage of $156 million over a two year period. (LIJ)

2/26/80 "N-Accident Cost Area $9 Million."
A NRC report estimates that the TMI accident caused net losses of $9 million within a 15 mile radius of the TMI plant. (HP)

2/27/80 "Monitoring: U.S. Official Says TMI's is Sufficient."
A US Public Health Official says that the federal agencies are doing an adequate job of monitoring radiation from TMI. (HP)

"Eleven State Lawmakers Urge Probe on Credibility at TMI."
Eleven state legislators send a letter to Pres. Carter asking that a top level person come to the area to assess the credibility problem of TMI monitoring and reporting. (HP)

"Releases of Gases at TMI Urged."
The final report of the Governor's Commission on TMI has been released. Among recommendations are expeditious cleanup of Unit II and controlled release of gases. (HP, LIJ, and NYT)

"NRC Ready to Lift License Moratorium."
For the first time since the TMI incident, the NRC is ready to begin licensing N power plants. (HP)

"14 Members on TMI Panel."
The article lists the names and affiliations of the 14 members of the Governor's Commission on TMI. (HP)

"Concern Over Growth of Area Following Accident."
The article reports that a study on the growth of the area near TMI found short

term disruption but no long term effects. (MPJ; also 3/1/80 LIJ)

2/27/80 "Council Questions Effect of TMI Resolution."
A 7/79 Middletown Boro Council resolution against the restart of Unit I is not being seriously considered which upsets council members. (MPJ)

"Speakers and Film Slated at PANE."
A March meeting program is announced. (MPJ)

"Special Force Will View Clean Up Operations at TMI."
The NRC establishes a special task force to monitor the cleanup progress at TMI. (MPJ)

2/28/80 "4 Johnstown Groups Seek OK to Join Met Ed Hearings."
Johnstown, Pa. business interest groups asked the PUC for permission to participate in hearings on matters involving Met Ed and the TMI accident. Among issues are should Unit I continue to be part of Met Ed's electricity rate base and the continuance of Met Ed's license. (HP)

Editorial: "State TMI Report."
The editorial discusses the report of the Governor's Commission on TMI. It says the report fails to take an overview of the problems associated with N power which have been delineated by the TMI accident. It sees the commissions recommendations leading to taxpayers subsidizing the hidden costs of N power. It states that the Commission was more intent on preserving N technology rather than protecting the public. (HP)

2/29/80 "Nuclear Panel to License Tennessee Power Plant Test."
NRC votes to block discharges of radioactive waste water from damaged TMI into the Susquehanna River. (NYT and HP)

2/29/80 "US Aid for TMI 'Unlikely.'"
Public support for Met Ed aid for TMI from the federal government is unlikely states two Met ed officials. (LIJ)

3/4/80 "Nuclear Industry Figures Labeled 'Misleading.'"
The EAF says a N industry report stating that N energy costs less to produce electricity than coal-fired power contained misleading statistics. (HP)

"N-Accident Reporting Bill Advances."
The Senate Consumer Affairs Committee approved a bill to require immediate reporting to state and local officials of any accident or equipment failure at a N facility that results in the release or potential release of radioactive material. (HP)

"N-Mishap Evacuation Plan to Commissioners."
The Cumberland County Office of Emergency Preparedness is nearing completion of a evacuation plan for the County Commissioner's review. (HP)

"NRC Pledge Venting Decision by Friday."
The NRC staff will complete an EIS and make their recommendations on TMI venting. (LIJ)

3/5/80 "Mayors Mull Over Lesson From TMI."
The mayors of the six communities surrounding TMI met to discuss issues arising from the accident. (LIJ)

"Anti-Nuclear Crowd Schedules TMI Rally."
With rock star Linda Ronstadt and Ralph Nader joining them, anti-N groups are planning a four day rally to mark the first anniversary of the TMI accident. (LIJ and MPJ)

3/5/80 "State, County Duos Meet."
Pa. aides met with Dauphin County Commissioners after their attack on the latest site report on the TMI accident. Points of dispute between Pa. and the county are discussed. (HP)

"Two Testify; Met Ed Customers Should Not Pay Costs of Unit I."
A report on the testimony on the Met Ed rate hike request is presented. Witnesses for the consumer advocate tell the PUC that consumers should not pay for the idle Unit I reactor. (HP)

"Unit II Shift Cost Put at $1.4 Billion."
GPU reports that Unit II could be converted to a coal or gas-fired plant at the cost of $1.4-$1.7 billion. The company says conversion is 5-10 times as costly and comsumes twice as much time as restoring the plant's N generating capacity. (HP; also 3/6/80 LIJ, and 3/12/80 MPJ)

3/6/80 "NRC: Clean Up TMI or Court Disaster."
Hendrie stated that the NRC is taking too long to decide the cleanup of TMI. He states that the "commissioners have been sitting on our duffs and fiddling, while the chances of serious radiation releases from the crippled plant increases with the deterioration of plant safety equipment." (LIJ and HP)

"Entry Into Unit 2 Airlock to be Initiated Next Week."
The procedures leading to entry into Unit II next week are being initiated. (HP)

"NRC Moving to Accelerate Cleanup at TMI."
The article discusses the discontent over the pace of the TMI cleanup. It outlines the steps being taken to accelerate the cleanup of TMI. (HP)

3/7/80 "GPU Says Delays in TMI Cleanup Courting Disaster."
GPU Pres. warns the NRC that delays in cleaning up Unit II could lead to a serious accident at the plant. The NRC Probabilistic Analysis Group prepared a response on the possible offsite consequences of possible future accidents at TMI and chances of further meltdown as slight. (HP)

"Conference Slated on Lessons of TMI."
The Continuing Education Division of PSU Capitol Campus has arranged a conference on lessons for 3/28/80. Representatives of media, emergency evacuation, Mayor Reid, and Anne Trunk will be panelists. (LIJ)

3/8/80 "Cover-up at TMI Again is Doubted."
Rogovin reports that confusion and incompetence rather than cover-up led to the withholding of TMI information at the time of the accident. (HP; also 3/12/80 LIJ and NYT)

"N-Waster Transit Prohibition Lifts."
The NRC lifts a ban on the shipment of low level radioactive wastes from TMI. It also reports on NRC inspectors who failed to act 'quickly' and 'prudently' to halt a Met Ed sampling procedure that allowed the release of krypton. (HP)

3/11/80 "TMI Staff Begins Recovery Process."
Signaling the start of recovery efforts at TMI, technicians flipped a switch Monday to slowly 'purge' radioactive gas from a plant airlock. (LIJ and NYT)

3/12/80 "City Director Evaluates Radiation Detection Unit."
Radiation detecting equipment for Lancaster's water supply will not detect isotope tritium states J. P. Angstadt, the city's Public Works Director. Tritium is expected to be the most common isotope

found if the decontaminated water is released into the Susquehanna River. He also said, "despite this limitation, the equipment will provide the city with valuable capabilities it does not possess." (LIJ)

3/12/80 "Met Ed on TMI: TMI is Money."
Arnold states costs will be high in terms of dollars and delay if the NRC refuses to allow the release of radioactive gas from TMI Unit II. (LIJ)

"NRC Considers Another Fine Against TMI."
The NRC has formed a staff group to determine whether the operator of the TMI reactor should be fined for not promptly informing the federal government about the full seriousness of the accident. (LIJ)

"Residents Concerned About Krypton Venting."
Middletown school officials indicated that a substantial number of parents are requesting to have schools closed during the venting. (MPJ; also 3/13/80 HP, and LIJ)

3/13/80 "NRC Staff Urges Venting of Radioactive TMI Gas."
NRC staff urges venting as it would take only 60 days and is safe. Other safe methods would take longer than a year. (LIJ)

"Nuclear Fallout Units May Stimulate Goodwill."
It is proposed that boroughs such as Goldsboro and Middletown be given their own radiation monitoring equipment. The NRC would train people to read the monitors and the federal DOE would pay for the monitors. (LIJ)

3/14/80 "Airlock Purge Done at TMI: Entry is Made." Radioactive gas was purged from the interior of Unit II and technicians entered it for the first time since the 3/28/79 accident. (HP, LIJ, and NYT)

3/16/80 "Inactive Reactors: One Year's Toll of Three Mile Island." An article reviewing setbacks the N power industry has sustained in the year since the accident at TMI. (NYT)

3/18/80 "Nuclear Power Backers and Critics to Mark 3 Mile Island Anniversary." Supporters and critics of N power plan demonstrations, news conferences, candlelight vigils, and other activities to mark the first anniversary of the TMI accident. (NYT)

3/19/80 "Nuclear Watch Panel Created." Pres. Carter signs an executive order establishing the NSOC which was recommended by the Kemeny Commission. (NYT)

"Public Hearing on Krypton Gas Venting Tonight." A hearing will be held by the NRC staff to explain its recommendation for the venting of Krypton gas. (MPJ)

"Special Resolution to Deal with Gas Venting will be Drawn Up." The Middletown Boro Countil will ask for a postponment of the venting of the Krypton. (MPJ)

"NRC Acts to Assure Immediate Reporting of Plant Accidents." New procedures are approved by the NRC for immediate reporting of plant accidents. (MPJ)

Editorial: "Krypton Legacy." The editorial argues for the use of lay people in monitoring TMI releases. It questions the worth of governmental assurances on the safety of Krypton doses. (HP)

3/20/80 "TMI Neighbors Tell NRC: We're Sick, We Are Tired, We Are Angry."
Four hundred people attended a public meeting in Middletown for the purpose of telling the NRC to explain the pro and cons of various proposals to get rid of the radioactive gas. A common theme expressed is that the residents are sick, tired, and angry about the TMI cleanup. (LIJ)

"Expert Says TMI Doomed."
The TMI Unit II is unlikely to reopen in light of mounting costs and public pressure, stated Dr. R. Parente. He is a management consultant associated with Theodore Barry and Associates, a firm hired by GPU. (LIJ)

3/21/80 "Met Ed's Venting Stirs Up a Storm at Elizabethtown."
Thursday night in Elizabethtown, officials of Met Ed felt the after-shocks of the emotional quake that hit Middletown earlier at a public meeting. (LIJ and NYT; also 3/22/80 HP)

"TMI Venting Due Study."
Stephen Reed, Dauphin County Commissioner and Pa. Rep., feels that federal funding will be provided for Harrisburg area citizens to assess the TMI venting. (HP; also 3/26/80 MPJ)

3/22/80 "Walker to be Present if TMI is Vented."
Cong. Walker will be present and he wants the entire Board of Directors of GPU to be present if TMI is vented. (LIJ)

"Venting of TMI Gas Could Cause Violence, NRC Told."
Civic leaders from Central Pa. warned the NRC on Friday that rioting could erupt if the agency allows the venting of Krypton to occur. (LIJ)

3/24/80 "TMI Effect on World Weighed."
An examination of N energy worldwide in light of the TMI accident is discussed. (HP)

"Trying to Clean Up Three Mile Island."
Future plans to decontaminate the crippled reactor at TMI are outlined. The article explains in general terms the first attempt at entering the building which houses the reactor. A tentative time schedule for the draining, scrub-down, and dismantling of the reactor is provided. If Met Ed goes bankrupt due to the exorbitant cost of the cleanup, or if its operating certificate is revoked, no one knows who will pay the cost of cleanup. (NW)

"Stress of TMI Restart May Be Considered."
The article reports on the mounting pressure on the NRC to permit consideration of public mental stress in hearings on the restart of TMI Unit I. Psychologists, Dr. J. Kalen and Dr. E. Martin, from HMC, are cited in support of this premise. PANE of Middletown has argued for including mental stress as a factor. Met Ed and the NRC staff have argued that stress cannot be 'quantified' and is not admissible by law in a licensing hearing. (LIJ)

3/25/80 "U.S. Wants Pa. Money Spent on TMI: Pa. Wants U.S. Money."
In hearings before the Senate N Regulation Committee, Sen. Hart urged that Pa. take a more active role in calming resident's fears. At Pa. PUC hearings, the commissioners requested federal funds to aid Pa. in the cleanup of TMI. (LIJ and HP)

"TMI Activists Plan Party to Mark First Anniversary."
The 3/28 Coalition announced a series of events over the weekend to commemorate the anniversary of the N accident at TMI.

The accident is an example of what can go wrong with N power in the US and anti-N leaders want people to remember it. (LIJ)

3/25/80 "Gov. Thornburgh Meets With 50 TMI Protestors."
A group of 50 demonstrators protested outside of the Pa. Governor's mansion. They were in opposition to venting of Krypton from TMI. (HP)

3/26/80 "TMI Overblown, Utility Study Says."
An utility industry study suggests that federal investigators have misled the public to believe there was a near disaster at TMI. (HP)

"N-Notification Measure Clears State Senate."
A bill requiring N plant operators to notify local authorities by the fastest means possible of an alert, site emergency, or general emergency at their plants cleared the Pa. Senate. (HP)

"No-Nukes Cancel Forum Concert."
A concert commemorating the first anniversary of the TMI accident has been cancelled. A smaller concert will be held. Other anniversary events are also listed. (HP)

"No-Strike Agreement is Signed at Disabled Nuclear Power Plant."
Workers at TMI agree in contract not to strike while the disabled plant is being rehabilitated. (NYT)

"Utility Suing Supplier Over Three Mile Island."
GPU sues B&W for $500 million in damages it has already incurred as a result of the 3/28/79 accident. (NYT)

"Anniversary Edition: TMI: A Year Later."
This is a series of stories about: a survey favoring closing of TMI, the GPU

suit against B&W, and a review of the past year in relation to TMI. Workers and people's dilemmas are emphasized. Plans for a 1st anniversary rally are also highlighted. (MPJ)

3/26/80 "Snyder to Head TMI Clean Up."
The NRC announced Monday that it appointed Dr. B.J. Snyder to head the federal agency's cleanup of the crippled Unit II reactor at TMI. (LIJ; also 4/2/80 MPJ)

"False Nuke Power Costs Under Attack."
The true costs of N power have been ignored to make it appear economical. This conclusion was presented to the PUC by Dr. Wm. Belmont. He appeared as a witness for the TMIA. The areas of insurance and spent fuel disposal were given as examples. (LIJ)

"No Problems Seen at TMI Anniversary."
A Pa. state police spokesman states that no trouble at the TMI anniversary vigil is expected. (LIJ)

3/27/80 "TMI: A Look at Then and Now."
Two stories dealing with the TMI accident are presented. The first one reviews the accident and the second assesses the current status. (LIJ)

Editorial: "TMI Crisis and One Year."
The editor hopes that by 3/81 the NRC will decide that TMI will not resume operation as a N power plant. (LIJ)

"Researchers Finding Anxiety in the Area Near 3 Mile Island."
Researchers have found signs of possible chronic stress among area residents as part of the mental health effects of TMI. (NYT)

"Radiation Items Termed 'Useless.'"
The NRC states that the surplus CD materials used by some local residents to monitor radiation are obsolete. (HP)

3/27/80 "Sample Posting Planned."
The NRC agrees to post radiological sample readings of offsite air around Unit II daily at their Middletown office. (HP)

3/28/80 "Year After Accident, Safety of Reactors is Still in Dispute."
An article on the continuing debate of safety of N reactors one year after the TMI accident. (NYT)

"Officials Seem to Favor Venting Gas at 3 Mile Island."
Denton supports venting of Krypton into the atmosphere. People near TMI oppose venting. Thornburgh just stops short of endorsing the venting approach. (NYT and LIJ)

"Venting Remains Choice of NRC, Reporters Told."
The NRC has decided to vent Krypton from Unit II as the most feasible approach to decontamination. A problem in selling the idea to the public is expected. (HP)

"Panel to Study Thyroid Ills."
A special study of infants born in Pa. during the last year who have hypothyroidism has begun. The study seeks to determine the impact of TMI on this problem. (HP)

"Activities to Mark TMI Anniversary."
The story highlights how a variety of activities are being planned in different parts of the country to mark the first anniversary of the accident at TMI. (LIJ; also 3/29/80 NYT)

"Anti-Nukes Join to Raise Money."
All the anti-N groups in Central Pa. have joined together to raise money for their various legal battles. Beverly Hess of SVA made the announcement and the group is called TMI Legal Fund. The six groups represented are: 1) SVA, 2) TMIA, 3) PANE,

4) Angry, 5) ECNP, and 6) Newberry's TMI Steering Committee. (LIJ)

3/28/80 "Psychiatrists Fear Chronic TMI Stress." Fourteen studies have indicated that the major health effect of the TMI accident is in the area 'of mental stress.' New data leads a psychiatrist to wonder if the stress hasn't already become chronic. (LIJ)

"Gov. Doubts Safety of Venting TMI Gas." At a press conference, Gov. Thornburgh announced, "he is not yet convinced that it is safe to vent radioactive gas from the disabled TMI plant." He will consult with Robert Pollard of the UCS. The NRC staff, radiation experts at Pa. DER, as well as Lt. Gov. Scranton have supported the venting proposal. (LIJ)

3/29/80 "TMI is Never Out of Her Sight, Mind." A human interest story with Fran Cain concerning her feelings about TMI. She is a member of the Londonderry Concerned Citizens. She said, " I feel like it's taken 20 years out of my life. It's the mental anguish, that's the thing." (LIJ)

"Rain Fails to Stop TMI Marchers." A story about an evening march in downtown Harrisburg to protest TMI is described. (LIJ)

"Nuclear Debate Ends Inconclusively." Public TV debate titled "The National N Debate" was held in Harrisburg. Dr. N. Rasmussen contended that N plants are structurally safe. Dr. H. Kendall, also of MIT, said they aren't. Dr. H. Caldicott is concerned over radiation hazard to the public health, whereas Dr. R. Linneman argued that radiation from TV is greater than that released by the accident. Dr. D.D. Rossin, engineer with Commonwealth Edison Company and Dr. V. Taylor, energy consultant, disagreed over the economic future of N power. Finally, Cong. Corcoran supported N power, whereas Cong. Markey opposed it. (LIJ)

3/29/80 "New TMI Poll Released."
About half of the people living within 15 miles of the crippled TMI N plant oppose the plans to vent radioactive gas from the reactor containment building. The survey was conducted by Dr. D. Kraybill of SRC. (LIJ)

"McGovern, in York, Calls for TMI Vote."
Speaking at York College of Pa., Sen. George McGovern of South Dakota said the public in the area of TMI should be allowed to vote by referendum on the future of the plant. (HP)

"Hiroshima Blast Memories Vivid."
A survivor of the atomic bomb blast in Hiroshima participated in first anniversary protest of TMI. She recalls memories of the blast and asks the audience to join the fight against N power. A Buddhist monk from Japan says public pressure is being brought on the Japanese government to reexamine its N power policy as a result of the TMI accident. (HP)

"TMI Anniversary Crowd Stern."
The article details some of the events connected with the first anniversary protest of TMI. (HP)

"Real Estate Unaffected, Group Says."
A Greater Harrisburg Realtors Study says that the property values within a 20 mile radius were not affected by the TMI incident. (HP)

"Bond Ratings Slashed."
Moody's Investors reduced their rating of bonds held by GPU saying it considers the issue speculative. (HP)

3/29/80 "N-Power is Called 'Evil.'"
A report on an interreligious service commemorating the first anniversary of the TMI accident. (HP)

"TMI Cleanup Called Paramount."
Representatives of the UCS met with Gov. Thornburgh. They discussed Krypton venting and the need to decontaminate Unit II. (HP)

3/31/80 "Area Anti-Nuke Protest TMI."
This story highlights an anti-N rally with a crowd of 7,000 with Dick Gregory, Pete Seeger, and Linda Ronstadt featured. (LIJ)

"Lancaster Urges Shutdown of TMI."
Betty Tompkins of SVA spoke before a group of D.C. area anti-N protestors. She called for the shutdown of TMI. (HP)

4/1/80 "County Passes Resulution Against K-85 Venting."
Dauphin County Commissioners pass a resolution opposing Met Ed and the NRC plans to vent Krypton from TMI. (HP)

"TMI Area Baby Death Data Eyed."
The PDH began a review of infant death statistics in the TMI area for the first six months after the accident. (HP)

"Middletown Meeting with NRC is Peaceful."
Middletown citizens conducted themselves in an orderly fashion as they met with NRC representatives to discuss venting. (HP)

4/2/80 "Pro-Nukes Break Silence."
A local realtor, Jack Stolz, spoke in favor of N energy at the Middletown Boro Council meeting. (MPJ)

"Future of Nuclear Power in Pa. is Indeed Cloudy."
Highlights of a story from a DER publication, *Econotes*, which reviews the current status of the N power industry in Pa. is presented. (MPJ)

4/2/80 "GPU Offers Conservation Plan."
GPU proposes an electricity conservation and reallocation plan. It projects a $1.2 billion savings to customers over the next 30 years if the plan is adopted. (HP)

"DER to Expand TMI Monitoring."
Residents from communities around TMI will be trained to use radiation monitoring equipment. (HP and LIJ)

"State Checks Baby Deaths in TMI Area."
Deputy Secretary Dr. D. Reid of the PDH indicated that the deaths within 10 miles of TMI during the first six months following the accident appear to be higher than normal. Further examination of the data is being planned. (LIJ)

4/3/80 "Mistake Made in TMI Death Survey."
Despite a recent jump, deaths around TMI do not show any significant difference from the Pa. rate according to PDH Secretary A. Muller. (LIJ, HP, and NYT)

4/4/80 "Lebanon Opposes N-Venting."
The Lebanon County Board of Commissioners passed a resolution opposing the venting of Krypton gas at TMI. (HP)

"Unit II Entry Denied."
The NRC denies Met Ed's request to send technicians into the Unit II containment building. The safety of technicians must first be ensured. (HP)

"TMI Benefits: Safety Focus."
A Congressional subcommittee views the TMI accident as raising the consciousness level of the N power industry and not as a threat to public health. (HP and NYT)

4/5/80 "Griffith Hits Venting Plan."
Mayor Donald N. Griffith of Lebanon urges the NRC to consider his opposition to the TMI venting. (HP)

4/5/80 "Safety Changes Proposed for Nuclear Power Plants."
The NRC staff says that it is not necessary to close B&W designed reactors but recommends a series of improvements to make them safer. (NYT)

4/7/80 "UCS Agrees to Study Gas Venting at TMI."
Gov. Thornburgh announced that the UCS has agreed to study the plan to vent radioactive gases from the TMI plant into the atmosphere. (LIJ)

"Three Mile Island's Legacy."
The emotional trauma experienced by residents living near TMI in the aftermath of the plant's accident is reviewed. The report tells of the protest rallies organized by Middletown area residents who claim to be keeping in mind their children's best interests. The people blame farm stock still births and other birth defects on radiation poisoning, although a Middletown veterinarian affirms that the cases he treated were caused by 'common diseases' or 'normal nutritional deficiencies.' In an effort to boost the popularity of N energy, the article reports that much money has been spent on improving the safety factors of power plants. The NRC is said to have acted to revamp the Commission's image. However, the article points out that in spite of these changes, N power's bad publicity is inescapable. (NW)

4/8/80 "Radiation Seeps Into TMI Walls; Leaks From Pipes, Not Disabled Unit."
Traces of radiation found in the test walls at TMI may indicate that the contaminated water has leaked into the ground, Met Ed officials reported. (LIJ and HP)

"Radiation Watcher Speaks Up."
A human interest story about a volunteer being trained to be a citizen monitor of radiation. (LIJ and HP)

4/8/80 "Nuclear Panel Backs Plans for 3 Mile Island Releases."
The NRC approved procedures for the release of low levels of radiation but still is undecided regarding the purging issue. (NYT)

4/9/80 "Shorter Venting Time Proposed by NRC."
The rapid purging of Krypton gas is recommended to reduce stress. (MPJ)

"Conservationists Vote Down Nuclear Plants."
A story reports the results of a membership survey of the National Wildlife Federation in which a majority opposes N power 51% to 40%. In 1978, a survey of the membership found 41% opposed to N power. (MPJ)

"Commissioners Pass Resolution that Opposes Krypton Venting."
Dauphin County Commissioners act to oppose Krypton venting. (MPJ)

"Survey Shows 88% Oppose Krypton Venting."
Three hundred residents were surveyed by a social psychology class at HACC and 88% are opposed to Krypton venting. (MPJ)

"Anti Nuke Groups Tell Thornburgh to 'Take a Stand.'"
Anti-N groups from the TMI area met for over 90 minutes with Gov. Thornburgh urging him to take a more active role in the plant cleanup. He is criticized for sitting on the fence. (LIJ and HP)

"Drilling at TMI Ordered by NRC."
The NRC asks Met Ed to do more drilling to pinpoint the source of leaking radiation. (HP)

"TMI-Related Death Chance Called Slim."
Persons living within 50 miles of TMI are said to run a 'one in a million' risk of fatal health problems caused by radiation released during the TMI accident. (HP)

4/10/80 "Vent the Krypton, Cumberland Officials Say."
Harold Denton made a surprise visit to a meeting of Cumberland County officials. Cumberland officials say vent the Krypton. (HP)

"TMI Neighbors Still in Shock and Fearful."
People living near the TMI N power plant remain in shock and in a potentially explosive mood more than a year after the accident, stated Anne D. Trunk. She addressed a NY Academy of Sciences Conference on TMI. Her thesis was supported by papers by Dr. C.B. Flynn and Dr. B.S. Dohrenwend. (LIJ)

"Morris Finds TMI on Denton's Menu."
Denton assured Mayor Morris of Lancaster that city and county officials would be fully briefed on plans to vent the Krypton gas at TMI. (LIJ)

4/11/80 "NRC Proposes $100,000 Fine for Designer of TMI Reactor."
A $100,000 fine is being proposed for B&W because they allegedly failed to provide safety information which might have prevented or lessened last year's accident. (NYT and HP)

"EPA to Handle TMI Monitoring."
The EPA has been designated as the primary federal agency to monitor offsite radiation levels around TMI. (HP)

Editorial: "Bad Advice and Poor Taste."
The editorial criticizes Cumberland County officials for supporting Krypton venting. (HP)

"NRC Will Issue N-Plant License."
A limited operating license will be issued to the North Anna Va. Unit II. This is only the second new license since the TMI accident. (HP)

4/11/80 "Officials to Write Carter on TMI Costs." Lebanon County Commissioners voted to send a letter to Pres. Carter complaining about the increased utility costs in the wake of the TMI N accident. (LIJ)

4/12/80 "TMI No Longer a Point in the Political Game."
A week before the Pa. presidential primaries, the article analyzes the impact of TMI as a national political issue and concludes that it is having minimal impact. (LIJ)

"Morris and Huber Favor Venting of Gas at TMI."
Mayor Morris and Lancaster County Commissioners Huber and Boyer are in favor of venting the Krypton gas. Lancaster County Commissioner Mowery and Mayor Griffith oppose the venting method. (LIJ)

"Trades Council Favors Venting."
The business manager of the Harrisburg and Central Pa. Building and Construction Trades Council says the organization supports Krypton venting, the TMI clean-up, and putting Unit I back on line. (HP)

"TMIA Raps Myers."
Commissioner Myers is criticized by TMIA for his support of the Krypton venting. (HP)

4/14/80 "N-Power Stands Aired."
Nine of the 12 Pa. candidates for US Senate advocate a moratorium on new N power plants. (HP)

Editorial: "Kendall Panel."
The editorial lauds the appointment of Dr. Henry Kendall to head a panel to investigate the venting of Krypton. (HP)

4/15/80 "Some Post-TMI Safety Rules Queried."
The ACRS of NRC warns that some newly mandated N power plant modifications may have created new hazards. (HP)

4/15/80 "Dauphin County Briefed by Met Ed."
Met Ed officials brief Dauphin County Commissioners on plans to vent Krypton. (HP)

"Cumberland to Ask About Venting."
A Cumberland County Commissioner seeks to arrange a public meeting on proposed venting and wants Denton to attend. (HP)

"TMI to Drill 4 More Wells in Leak Search."
Plant operator continues to seek the sources of excess radioactivity in ground water. It has been more than one week and they are still unable to locate the source. (LIJ; also 4/16/80 MPJ)

"Governor Seeks Delay of TMI Plan."
Gov. Thornburgh asks for a delay of the decision regarding venting for one month so that he would have the benefit of the results of the UCS study. (LIJ)

"3 Mile Island: No Health Impact Found."
Federal and PDH officials find no evidence that radiation released during the accident at TMI caused any damage to unborn children that resulted in birth defects or deaths during infancy. (NYT)

"Physicist Calls for Selective Absorption Over Venting."
Gerald Pollack presents plans to the NRC recommending the selective absorption approach for the release of the Krypton gas. (MPJ)

"LD Board Asks for Krypton Venting Delay."
The Lower Dauphin School Board asks that venting be delayed until summer recess. (MPJ)

"PANE Will Report on Legal Fund."
At its next meeting, PANE will report on its legal fund. (MPJ)

4/16/80 "Phantom Taxes Cost Met Ed Customers $102."
A study by the EAF of D.C. indicates that each Met Ed customer subsidized the company's expansion to the amount of $102 in 1978 in phantom taxes. TMIA is using this study to suggest that tax laws encourage expansion of N plants. (LIJ)

"Briefs Assert Met Ed Should Remain Viable."
Met Ed, the PUC legal staff, and the Pa. consumer advocate filed briefs with the PUC indicating that Met Ed should not lose its license. They disagree as to what is needed to keep the utility financially viable. (LIJ)

"Anti-Nukes Say Venting TMI Means 10 Cancer Deaths."
The TMI Legal Fund issued a 140 page report strongly opposing the Krypton venting plan. They contend that no level of radiation release is safe and the groups urge a "selected absorption system." (LIJ and HP)

4/17/80 "TMI Stand Requested by Mayors."
The governor is urged to take a stand by five mayors. Morris, Doutrich, and Reighard support the venting. Griffith opposes and Marshall is still undecided. (LIJ)

"Teddy Says TMI Utility Should Pay."
Campaigning in the presidential primary in Harrisburg, Sen. Kennedy states that there should be no venting of the gas Krypton and the utility should bear the costs of the cleanup. (LIJ)

"TMI Clean Up Pictorial Display Added."
A display is added to the Observation Center at TMI. It is also noted that there were 53,000 visitors since 7/79 whereas 15,000 visited per year prior to the accident. (LIJ)

4/17/80 "Governor Disputes TMI Opinion of Anti-Nukes."
The governor terms 'ludicrious' and 'irresponsible' the findings of ten cancer deaths made by the TMI Legal Fund regarding venting. No estimate of that kind of hazard is made by any other group. (LIJ)

"Stress of TMI Residents Rises."
As many as 93,000 people living near TMI have shown symptoms of mental stress because of the accident a PDH report concludes. Examples of stress include alcohol, smoking, use of tranquilizers, and sleeping pills. (LIJ and HP; also 4/18/80 NYT)

"NRC Insisting on Proof TMI was Negligent."
Persons who filed claims over last year's TMI N power plant accident will have to prove negligence to win their claims. (LIJ)

"Utility Power to Operate During Accident Urged."
ACRS of NRC recommends that the operation of a N power plant remain in the hands of the operating utility during an accident. (HP)

"NRC Decides TMI Accident Not 'Extraordinary.'"
The NRC claims that TMI accident was not extraordinary enough to be declared an extraordinary N event under federal law. (HP)

4/18/80 "LACI Board Backs Venting of Krypton Gas."
The LACI as well as Pa. CC supports venting of the Krypton gas. (LIJ)

"Ex-TMI Worker Claims Tampering with Tests."
Stello states that the Justice Department has been asked to investigate allegations made by H.W. Hartman, Jr. of tampering with test data on leaking prior to the TMI accident. (LIJ and NYT)

4/18/80 "Long Distress Found Over Atom Accident." A study finds that anxiety resulting from TMI still remains and that thirteen percent of 37,000 people living within 5 miles of TMI have become anti-N activists. (NYT)

4/19/80 "Birth Defects Not Linked to TMI." The PHD reported no link between the TMI accident and 34 cases of thyroid birth defects. Findings were the result of a special investigation panel and the names of the members with their affiliations are presented. (LIJ)

4/21/80 "Carter Pledges Safe TMI Clean Up Action." Pres. Carter pledges a safe TMI cleanup during an interview with a local reporter. (LIJ)

4/22/80 "NRC Authorizes Subpoena of TMI Employees." The NRC authorizes a subpoena in relation to allegations of excessive, unreported leaks from the reactor before last year's accident. (LIJ and HP)

"Alternative to Venting Pushed." Cong. Ertel asks the NRC to consider selective absorption system as an alternative to the Krypton venting in the TMI cleanup. (HP)

"Cause Unknown in Blaze That Guts TMI Area Trailer." An office trailer that stood 20 feet from the TMI Observation Center was gutted by a fire of undetermined origin. (HP)

4/23/80 "2 Will Enter TMI's Unit II on Thursday." The NRC approves two Met Ed technicians entering Unit II. The details of entry and possible hazards are discussed. (HP and LIJ)

"Citizen Monitors-Good Feelings." Volunteers are trained to monitor Krypton venting. (MPJ)

4/23/80 "Selective Absorption Start, Possible Within 60-90 Days."
An update of Prof. G. Pollack's analysis and recommendation for selective absorption to remove gas from TMI are presented. (MPJ)

"NRC Fines B&W $100,000 for Violations."
The NRC staff decides to fine B&W because of its failure to report safety information to the NRC. (MPJ)

4/24/80 "Met Ed Calls Off Probe Into 'Hot' TMI Building."
A probe into the TMI building is postponed due to failure to gain federal certification for breathing devices the men will wear. (LIJ)

4/25/80 "Gas Release Rumor Clouds TMI Clean-Up."
A mix-up in communication led to the belief that there was an immediate release of Krypton gas. (LIJ)

4/26/80 "Feds Think 'Small' in TMI Campaign."
Government officials are switching tactics from large public meetings to small home meetings to discuss proposed venting of Krypton Gas. (LIJ)

4/28/80 "Jersey Utility Seeks a Rate Rise of 20%."
JCPL is in financial difficulty because of its corporate connections to the damaged TMI and is expected to ask for a 20% rate hike. (NYT)

4/29/80 "Three Mile Island's Bees."
D.J. Lisk, Professor of Toxicology at Cornell University, says honey collected from beehives within 6 miles of TMI is free of radioactive contamination even though bees tend to magnify local contaminants in their honey. (NYT)

"Farm Group Asks Quick Removal of Gas."
The Pa. Farmers Association is urging that problems surrounding the removal of Krypton gas be quickly solved. (HP)

4/29/80 "Radiation Increase Unsolved."
A small radiation increase at TMI two weeks ago cannot be explained by the EPA. (HP and LIJ)

"Delay Krypton Venting, LD Asks."
The Lower Dauphin School Board again asks that venting of Krypton be delayed until the summer when schools are closed. Otherwise, they fear absenteeism. (HP)

"Borough Urged to Take Stand on Venting."
The Middletown Council Borough is urged to take a stand on the Krypton venting. (HP)

4/30/80 "Results of Radiation Monitoring."
EPA readings of radiation will be published as a public service. (MPJ)

"Area Schools Briefed on TMI."
A status report on Krypton venting is given by federal and Pa. agencies to area schools. (MPJ)

"Chamber of Commerce Supports Venting of TMI's Krypton Gas."
Pa. CC announces its support of venting of Krypton. (MPJ)

5/1/80 "GPU Net Halved; Reactor Ban Cited."
GPU reports that its 1st quarter net income fell 52% primarily due to revenue loss from the closing of the undamaged reactor at TMI. (NYT)

"House Members Approve TMI Citizen Committee."
The House Interior Committee unanimously recommended the creation of a special 15 member citizens advisory committee to oversee the cleanup of Unit I. (LIJ)

"Blame TMI's Operators, Not Reactor, Teller Says."
Teller, the father of the H-bomb, says operators, not reactors, are the blame for the TMI accident. (LIJ)

5/6/80 "TMI Again Prevents Dividends."
GPU withheld common stock dividends for the second quarter in a row. Shareholders criticized company officials. (HP and NYT)

"House Adopts N-Resolutions."
The Pa. State House approved 13 of the 15 resolutions offered by the Select Committee on the TMI accident. (HP)

"Radioactive Water Dumped in River."
The NRC has authorized Met Ed to dump water with a higher than normal amount of Tritium into the Susquehanna River. (HP; also 5/7/80 LIJ)

5/7/80 "Lancaster Eyes Met Ed Dumping."
The city of Lancaster is seeking to determine if the dumping of radioactive water from TMI test wells into the Susquehanna River is violating the agreement between the city and the NRC. (HP)

"Panel OKs Revised NRC Plan."
The House Government Operations Subcommittee endorsed Pres. Carter's reorganization plan for the NRC. (HP)

"Are the TMI Fears Unfounded?"
PDH Secretary Muller states that fears about TMI are unfounded and urges worries to be placed in proper contest. (HP)

"Royalton Unit Backs Vent Plan."
Royalton Borough Council adopts a resolution supporting venting of Krypton gas at TMI. (HP)

"Reactor Builder Gets Extension."
B&W wins a 20 day extension of appeal of $100,000 fine imposed by the NRC for failure to report safety information. (NYT)

"TMI Monitoring Network Results."
The 1st EPA readings are given in 18 locations. All readings are normal. (MPJ)

5/8/80 "Nuclear Oversight Panel Named."
Pres. Carter names Gov. B. Babbitt of Arizona to be Chairman of NSOC. (NYT)

"New Cumberland Session Erupts Over Venting."
The NCCANE condemned those Cumberland County Council members who signed a resolution favoring the release of Krypton gas from TMI. (HP)

"Scientist Against TMI Gas Release."
Speaking at a SVA meeting, Dr. B. Moholt opposes Krypton venting. (LIJ)

5/9/80 "Ertel Disputes NRC on TMI Clean Up Time Table."
Cong. Ertel states that selective absorption of radioactive gas could be built and tested within six months. NRC estimate is 2 years which is one reason why the NRC is in favor of venting. Cong. Ertel opposes venting. (LIJ and HP)

5/10/80 "Alternative to Venting Feasible-Not Better."
A description of how selective absorption system could be used to cleanup TMI is presented. It is an alternative to venting, not necessarily better. (HP)

"PUC Lets Met Ed Live, Grants Extra Revenue."
The PUC says Met Ed should continue to operate as a public utility. It suggests that Unit I be removed from the rate base, but that its energy rate charge be increased. (HP and LIJ)

Editorial: "Go Ahead for Met Ed."
The editorial criticizes the PUC's failure to revoke Met Ed's license. The editor suggests that this indicates a preference for restart of Unit I rather than protecting the public. (LIJ)

5/13/80 "Grand Jury Selected to Weigh TMI Case."
A federal grand jury is selected to hear testimony on the alleged safety violations at TMI prior to the accident. (LIJ and HP)

"'Silent Majority' Declares Pro-Nuke Stance to NRC."
An ad hoc group of Central Pa. residents presents a proventing view to the NRC. (HP)

5/14/80 "Bar 'Coalition' From Hearings NRC is Asked."
Met Ed lawyers have asked the NRC to bar the ECNP from hearings on the future of Unit I. The request is based on the ECNP's nonresponse to Met Ed's questions. (HP)

"Met Ed to Seek Another Rate Hike."
Met Ed will seek another rate increase. It is expected that the PEC will file for the same. (HP)

Editorial: "Second Chance."
The editorial discusses the PUC decision on the removal of Unit I from Met Ed's rate base and rate hike request. The editor sees this decision as least disruptive. (HP)

"Team Will Enter Unit 2 May 20th."
Entry into Unit II is rescheduled due to the failure to have the approval of the NIOSH concerning breathing apparatus of workers' uniforms. (MPJ; also 5/15/80 LIJ)

"TMI Monitoring Network Results."
All radiation readings are of normal background. (MPJ)

5/15/80 "Radioactive Water Pact Intact, Official Says."
Lancaster City official does not view the dumping of radioactive water from the TMI

test wells as violating the city and NRC pact on dumping water into the Susquehanna River. (LIJ)

5/15/80 "Mental, But Not Physical, Venting Effects Foreseen."
The UCS report on venting suggests that the process may have a mental health but not a physical impact. (HP and LIJ)

5/16/80 Editorial: "Krypton Venting."
The editorial states that the time has come for a NRC decision on venting of the Krypton gas. The editor discusses the UCS report. (LIJ)

"Venting Mental Anguish Noted by Area Physicians."
Area physicians conclude that venting is safe. They also state that mental anguish is due in a large part to communication failures. (LIJ)

5/17/80 "Governor Backs TMI Gas Venting."
Gov. Thornburgh concludes that the safeness of the Krypton venting plan as analyzed by UCS should reduce mental stress. Therefore, he favors this approach. (LIJ and HP)

"Met Ed Vs. PUC."
Met Ed protests the PUC decision to eliminate Unit I from the utility's rate base. (HP)

"NRC Bars Intervenors Fund Help."
The NRC refuses to pay legal expenses incurred by intervenors in hearings on the restart of Unit I. (HP)

5/20/80 "NRC Finishes Venting Study."
The NRC staff recommends venting of Krypton gas. (LIJ)

"Governor's Council Supports Nuclear Plants If Made Safe."
The Governor's Energy Council said N plants should be allowed to go on operating

if they satisfy federal safety and emergency standards. They also recommended that restart of TMI units be based primarily on safety, secondarily on costs. (HP)

5/20/80 "Infant Deaths Lower at Three Mile Island." PDH releases 1979 infant mortality figures showing lower deaths in the area around TMI than elsewhere. (NYT and HP)

5/21/80 "Jammed Door Bars Atom Plant Entry." The first human entry into containment building is delayed due to a jammed airlock door. (NYT, LIJ, and HP)

"Krypton 85 Gas Venting." Two articles are presented which cover the arguments for and against the Krypton venting. (MPJ)

"TMI Monitoring Network Results." All radiation readings are of normal background. (MPJ)

"B&W Agrees to Pay $100,000 Reactor Fine." Despite paying the $100,000 fine, B&W denied all claims of noncompliance by the NRC. (LIJ)

5/22/80 "TMI to Limit Next Entry Try." The article discusses the airlocked door that prevented manned entry into Unit I. It also details Met Ed's plans for entry. (HP and LIJ)

"City to Monitor Water Radiation by Next Month." The city of Lancaster will assume complete responsibility for monitoring water for radiation within the next 14 to 30 days. (LIJ)

5/23/80 "TMI's Plan for Venting is Defended." Testifying before the House Interior Subcommittee on Energy and the Environment, Arnold defended Met Ed's plan to vent radioactive gas to the atmosphere. (LIJ)

5/24/80 "PUC Gives TMI Right to Operate."
A final order is given by the PUC giving Met Ed the right to operate TMI. The order also removes the idle TMI reactor from the rate base charged by Met Ed. (LIJ and HP)

5/27/80 "Middletown Places Limits on Call for TMI Venting."
The Middletown Borough Council adopts a resolution calling for quick venting of Krypton from TMI, but under certain conditions. (HP)

5/28/80 "TMI Monitoring Network Results."
All radiation readings are of normal background. (MPJ)

"Residents View Film Dealing with TMI."
Two hundred people at a PANE meeting see the film, *We Are The Guinea Pigs*.

5/29/80 "TMI Stress May Reduce Births."
Joseph McFalls, Jr., a Temple University sociologist, believes psychic stress resulting from the N accident at TMI may cut down on the area's birth rate. (LIJ)

5/30/80 Editorial: "The Lingering Stress of TMI."
The editorial criticizes Prof. McFalls, Jr.'s statements on stress as they are based on supposition, not research. (LIJ)

5/31/80 "New Study Says TMI Mental Effects Only Minor."
Findings of a NIMH supported study indicate that the mental health effects of the TMI accident were minor. The article indicates who the researcher was and the samples used in the study. (LIJ and NYT)

6/3/80 "US Files Suit to Compel 5 TMI Workers to Testify."
The Justice Department is filing a suit against the 5 TMI employees who refused to testify last month after each of them was subpoenaed by the NRC. Names and

positions of the five employees are provided. (LIJ, HP, and NYT)

6/3/80 "NRC Backs Purging of Unit 2."
The NRC in a final recommendation called for the purging of Krypton from Unit II over a two month period. (HP and NYT)

"TMI Entry Seen Foiled by a Latch."
A safety latch rather than corrosion is speculated as the possible cause of the inability to open the airlocked door at Unit II a month ago. (HP)

6/4/80 "Group Accuse NRC of Duplicity in Venting Plan."
Anti-N groups charge that the NRC selectively solicited comments from pro-N groups in considering the venting of Krypton gas at TMI. (HP)

"Questions, Rumors About TMI?"
The MCNEC has established a 24 hour telephone system regarding TMI. (MPJ)

"TMI Monitoring Network Results."
All radiation readings are of normal background. (MPJ)

6/6/80 "NRC Supports Venting TMI in Straw Poll."
In a straw poll, 4 of the 5 NRC Commissioners favor venting. Staffers criticized Cong. Ertel's statements concerning the viability of charcoal or freon absorption of Krypton gas. (LIJ and HP)

6/10/80 "State Reports on TMI Steps."
Gov. Thornburgh presents a report to the House Select Committee on TMI listing steps the administration has taken in the wake of last year's N accident. Actions listed in regard to PEMA, DER, banking, and education. (LIJ)

6/11/80 "TMI Gets Go-Sign to Vent Krypton."
In an unanimous vote, the NRC gives permission to vent gas which will occur

in June. The TMI Legal Fund will not oppose venting due to limited financial assets. Doreen Snell of SVA states that the report of UCS has helped to calm stress even though she is still opposed. (LIJ, MPJ, HP, and NYT)

6/11/80 "TMI Monitoring Network Results." All radiation readings are of normal background. (MPJ)

Editorial: "Venting Foes Won't Contest NRC Decision." The editorial criticizes the NRC decision to vent the Krypton gas. (LIJ)

6/12/80 "TMI Evacuation Costs $8 Million." A report titled "Health-Related Economic Costs of the TMI Accident" is released by the PHD. The report states that about 150,000 people evacuated the TMI area during the two week period following the N accident. The evacuation costs reached $6.9 million, plus an additional $1.4 million in loss pay. (LIJ)

6/13/80 Editorial: "Venting Decision." The editorial discusses the NRC decision on venting. It agrees with local anti-N groups that there is no need to go to court to challenge the decision. It calls for the NRC and the N industry to work towards zero emmissions. (HP)

6/14/80 "TMI Venting Starts June 28." The TMI Krypton venting decision and process is discussed. (HP)

6/17/80 "N-Power Expansion is Favored in School Plan." Students at Cedar Cliff High School surveyed fellow students regarding N power. A majority favor expansion of N industry. (HP)

"Senate Send Bill to Pres. on N-Safety." A bill is sent to the Pres. which withholds N licenses unless emergency response plan for the applicant plant exists. (NYT)

6/17/80 "Advertisement: Public Notice by USNRC." The purge is to begin as early as 6/22/80. A report analyzing venting as well as alternatives is provided in "Final Environmental Assessment for Decontamination of the TMI Unit 2 Reactor Building Atmosphere." (LIJ)

"SVA Broadens Lawsuit Over Water at TMI." The SVA hopes to prevent the discharge of the decontaminated water at any time. The key issue is treatment and disposal of 850,000 gallons of radioactive water. (LIJ)

6/18/80 "SVA Says TMI's Clean Up Threatens Area: Files Suit." The article provides the documentation for the SVA law suit. A list of the 25 plaintiffs with their addresses is also provided. (LIJ and HP)

6/19/80 "Germans Challenge Gas Venting." Bernd Franke and Dieter Teufel reanalyzed the radiological assessments made by the NRC and Met Ed and conclude "venting of Krypton gas at the crippled TMI could expose nearby residents to cancer-causing levels of radiation." (LIJ, HP, and NYT)

"NRC Reversal Could Assist Plant Critics." The NRC has decided that Class 9 accidents must be a consideration when granting licenses for N plants. (HP)

Editorial: "SVA Earns Our Support." The editorial praises SVA for expanding its legal suit and for focusing on the Epicor II system as the important issue in the cleanup. (LIJ)

6/20/80 "Author Claims TMI Report was Misread." Franke claims federal officials misread his report. (LIJ)

6/20/80 "Met Ed: Worse of Gas Venting-No Health Hazard."
At a meeting for public officials, Arnold of Met Ed said that the worse accident imaginable during the venting does not pose a health hazard. (LIJ)

"Vent Details Shape Up."
The article reports on the technical details of the venting process. (HP; also 6/25/80 MPJ)

6/21/80 "Unit 2 Entry Curb Blamed on Jammed Safety Lock."
An aborted entry into Unit II on 5/20 is blamed on a jammed safety lock. (HP)

6/24/80 "Huber Checks Monitors as Venting Approaches."
In addition to the community monitors, Met Ed, DER, USEPA, USDOE, and the NRC will do their own monitoring. (LIJ)

6/25/80 "Anti-Nukes Press TMI Gas Fight."
Steven Sholly, PANE, and the Newberry Township TMI Steering Committee will go into court to prevent the release of Krypton until a public hearing is held. (LIJ, MPJ, and HP)

"60% Oppose, Not Approve TMI Reopening."
A study of residents residing within 5 miles of the plant indicates that 60% oppose the reopening of TMI. (MPJ)

"TMI Monitoring Network Results."
Radiation readings are of a normal background. (MPJ)

6/26/80 "Ruling Due on Venting Challenge."
A three judge panel will rule whether to hear oral arguments on the legality of the NRC venting order. (HP)

6/27/80 "TMI Venting May Start Without Key Hearing."
A three judge panel says there is no immediate relief for anti-N group seeking to fight purge of Krypton. (HP and NYT)

6/27/80 "Governor Restates Faith in Vent Safety." Gov. Thornburgh calls for calming of public anxiety relative to TMI venting. (HP)

6/28/80 "TMI Venting Begins 8 A.M., Few Evacuate." Venting will begin at 8:00 AM as lawyers failed to persuade a federal court to halt the scheduled venting. (LIJ and HP)

"Many Talk of Flight, But Few are Leaving." The human interest story highlights interviews with area residents. (HP)

"Arnold Family Not Afraid of TMI Fallout." Arnold of Met Ed and his family moved into a trailer near TMI as part of his campaign to indicate venting is safe. (HP)

Editorial: "It Was Just Another Unimportant Mistake."
The editorial states that the spilling of 10,000 gallons of radioactive water at Unit I indicates that TMI is still having plumbing problems and incompetent workers. (HP)

"3 Mile Island Venting Near, Hundreds Leave."
Hundreds of people leave the area for venting. A number of wives of the employees and officials of Met Ed join reporters at an observation point to view the venting. (NYT)

6/29/80 "Gas Releases Halted at Nuclear Plant." Releases of radioactive Krypton gas from the damaged reactor at TMI are halted almost as soon as they begin on 6/28, after the automatic radiation monitors gave an apparently false signal of danger. (NYT)

6/30/80 "Full-Scale Venting of Krypton Begins." The monitoring by government officials and representatives of citizens groups indicate that radiation readings are at normal levels. (NYT, HP, and LIJ)

7/1/80 "Venting is Rated Excellent."
The article reports that the venting is making excellent progress. (HP and NYT)

7/2/80 "Weather Benefits Venting."
Normal radiation levels are reported in the TMI area despite the venting of Krypton. (HP)

"TMI Calls Ease, 15% Gas Vented."
Fourteen thousand calls to hot line have been made with the number decreasing each day. Fifteen percent of the gas has been vented. (LIJ and NYT)

"Monitors View of the Krypton Venting."
The article describes the different values that can be attained on a radiation monitor. (MPJ)

"Carter Signs a New Law on Nuclear Plant Safety."
Pres. Carter signs into law a bill setting stricter safety standards for N power plants. (NYT)

7/3/80 "Senate Report on TMI Accident is Released."
The report concludes that the evacuation of the area should have started within hours due to uncertainty of the condition of the reactor core. (NYT and HP)

7/4/80 "Rusty Pin is Freed at TMI."
A rusty deadbolt that prevented entry into Unit II has been freed. Attempts to enter the building are projected for the future. (HP)

7/5/80 "Venting One-Third Completed."
More than one third of Krypton in Unit II is reported to have been vented. (HP and LIJ; also 7/6/80 NYT)

7/7/80 "Venting of TMI Gas Nears Halfway Mark."
Independent monitors reach the same conclusion as government monitors, that everything is within normal limits. (LIJ and HP)

7/8/80 "Venting Rate to be Upped."
A larger volume of Krypton purging is planned. Venting passes the 50% mark. (HP)

"Unite, TMI Area Customers Urged."
The Coalition of Concerned Consumers is seeking to unite all consumers in the GPU/Met Ed system on the rate hike issue. (HP; also 7/9/80 MPJ)

7/9/80 "U.S. Won't Aid Cleanup at TMI."
Pres. Carter denies any authority to offer direct financial aid to TMI for cleanup. (HP)

"Met Ed Speeds Venting of Gas."
The venting at TMI is now on an accelerated system. An alarm went off because of a faulty sensor, not due to radiation. (HP and MPJ)

"Krypton Venting is Not Harmful."
The NRC contends that the Heidelberg Report findings are erroneously high. Technical arguments are presented to dispute the Heidelberg allegations that Krypton venting is harmful. (MPJ)

"EPA Results Posted from Monitoring of Krypton 85."
EPA results are presented which indicate that no radiation above background was recorded. (MPJ; also 7/10/80 LIJ)

7/11/80 "Judge Upholds Law Suits by TMI Residents."
A federal judge ruled that all persons and businesses within a 25 mile radius of TMI may sue for economic losses suffered as a result of the 3/28/79 N accident. The judge did not permit claims due persons suffering physical and emotional injuries. (LIJ and HP)

"No Sign of Krypton Traced by Monitors."
Interviews with citizen monitors indicate no trace of Krypton noted. Ron Davis of SVA states that the water situation

presents problems and dangers that are much greater than the dangers presented by the Krypton. (LIJ)

7/11/80 "TMI Venting All But Over; Entry Looms." The Krypton venting is almost over. Plans proceed for the entry of a two man team into Unit II. (HP and NYT)

7/12/80 "NRC Reports Venting of Krypton is Completed on 7/11." The exposure of the population to radiation was no more than 4% of the officially permissible level. People are returning. (NYT and LIJ)

7/13/80 "Costs of Repairing A-Plant are Rising." Despite rising costs, Met Ed says the restoration of the damaged TMI N reactor should be completed by the end of '83. (NYT; also 7/14/80 HP)

7/15/80 "Accident to Put Agencies to Test." A simulated accident is planned for TMI. This will test Pa. and local responses and help secure federal approval for a statewide emergency plan. (HP and NYT; also 7/16/80 MPJ)

"NRC Finds Met Ed in Compliance for Unit 1 Restart." A NRC staff study indicates that Met Ed has met the short term and long term requirements for restart of Unit I. (MPJ)

"PANE Requests Ruling on Stress." PANE has asked the ASLB to rule that psychological stress should be included as one of the factors to be considered as part of the restart issue. (MPJ)

7/17/80 "TMI Drill Shows Pa. Emergency Plan OK." Henderson of PEMA indicates that the emergency drill was a success. (LIJ)

"Foul-Up at TMI." An analysis of the simulated accident at TMI is termed a disaster by some observers. (HP)

7/19/80 "Drill's Communications Weak."
PEMA officials now concede that a drill 'formerly termed successful' had communication problems. (HP, LIJ, and NYT)

7/24/80 "TMI Re-Entry."
The article details the first manned entry into TMI Unit II. (HP and LIJ; also 7/26/80 NYT)

7/25/80 "Met Ed Still Has Work To Do On Unit I."
The article lists the NRC requirements that have to be met before Met Ed can restart Unit I. (HP)

7/28/80 "Rule on Reactor Safety Adopted by Commission."
The NRC approves a new regulation that would require atomic reactors to be shutdown and would withhold a license from new plants if the NRC finds the emergency plan inadequate. (NYT)

7/30/80 "TMI Monitoring Network Results."
Radiation readings are of normal backgrounds. (MPJ)

"NRC Group Will Discuss Unit 1 Restart."
The ACRS will discuss Unit I restart. (MPJ)

8/2/80 "Venting Finished Early."
More gas was inside Unit II than expected. This caused a small purge of Krypton to be completed early. Small ventings will be necessary from time to time. (HP and LIJ)

8/6/80 "TMI Monitoring Network Results."
Radiation readings are of normal backgrounds. (MPJ)

8/8/80 "Met Ed Fined for Violating Shipping Rules."
The NRC fined Met Ed $9,000 for three violations of federal packing standards in the shipping of radioactive materials occurring in 2/80 and 3/80. (LIJ)

8/9/80 "TMI Accident Costs Tallied; $3.1 Billion."
The cleanup and replacement power costs are reported. The breakdown for cleanup costs is analyzed. (HP)

8/10/80 "Federal Study is Pessimistic in Fleeing Atom Accidents."
After a 6 week delay imposed by Pres. Carter, FEMA issues a gloomy report on the status of federal and state preparedness for another N accident like the one at TMI. The report says communication equipment, public alerting, and notification systems are not in order. (NYT)

8/12/80 "Re-Entry Set Friday at TMI."
A four man team is expected to engage in the second entry into Unit II since the accident. (HP)

8/13/80 "Met Ed Ordered: Give Data."
The NRC wants more data on the 6/27/80 leak of radioactive water on the containment building floor. They also want to know if maintenance was improperly delayed for undamaged Unit I. (HP and LIJ)

"Study: TMI Clean Up No Threat to River."
A draft version of a NRC EIS is ready for release. The study concludes that the cleanup of the crippled TMI N plant, including disposal of radioactive water, will not pose a significant threat to the environment. (LIJ; also 8/15/80 HP and NYT)

8/15/80 "5 TMI Employees Told to Testify Before NRC."
Five Met Ed employees who refused to testify about the accident were ordered to comply with subpoenas. (HP)

8/15/80 "Unit 2 Gas Vented: TMI Team Ready."
The article details the second manned entry into Unit II; more Krypton is vented. (HP; also 8/16/80 LIJ and NYT)

8/16/80 "Heat Overcomes 2 Who Entered TMI-2 Facility."
The article details the second manned entry into Unit II. Two of the technicians were overcome by heat and fatigue. (HP)

"Saturday Evac. Drill Cancelled."
Middletown and Royalton Boroughs planned a trial evacuation, but it was cancelled due to concern over insurance coverage and lack of cooperation between county and state officials. (HP; also 8/20/80 MPJ)

8/21/80 "Met Ed Case for $35 Million Weak."
A judge declares the use of extraordinary rate relief for the cleanup of TMI is not valid. (HP)

"Good Nuclear Engineering Grad's Hard to Find."
The TMI accident is offered as one reason for student disinterest in careers in N engineering. (HP)

8/23/80 "Gas Vented From TMI."
An unscheduled, but uneventful, venting of gas took place at TMI. (HP)

8/26/80 "Radiation at TMI Less Than Thought."
Radiation data gathered from inside the TMI plant leads the NRC to believe that cleanup will be less hazardous than expected. (HP; also 8/27/80 LIJ)

8/27/80 "Factors of Radiation Dosages, Stress, Waste Addressed."
The NRC staff has asked for public comment on their draft EIS which indicates that the disposal of the decontaminated water is not a threat to the environment. (MPJ)

8/28/80 "Delegation Vows Met Ed Bail Out Aid."
Pa. Senators are banning together to work on a federal subsidy bill to aid the TMI II cleanup. (HP and LIJ)

8/31/80 "Hundreds of Residents Left Three Mile Island Area."
The PDH reports that hundreds of residents within a 5 mile radius of TMI moved away permanently after the '79 accident. Some of them moved because of the emotional stress related to the accident. (NYT)

9/2/80 "Palmyra Votes to Write to Carter About TMI Power Cost Concerns."
The Palmyra Borough Council will write Pres. Carter expressing their concern over increased rates for Met Ed customers since the TMI accident. (HP)

9/3/80 "Public Comment Period is PANE's Concern."
The period of time for public comment on the EIS is too short says PANE. They are requesting expansion from 45 to 60 days. (MPJ; also 9/4/80 HP and NYT)

"NRC Upgrades Requirements for Nuke's Emergency Planning."
The NRC will require emergency response planning to be approved by the FEMA. (MPJ)

9/4/80 "State is Hot Over Route 11-15 as Road for Moving TMI Waste."
T. Gerusky of Pa. BRP feels that the use of 11-15 rather than route 81 is unsafe. Because of traffic lights and congestion, it is possible that a person could stand near a truck carrying disposable waste for three minutes. By NRC's estimate, that person would receive a radiation exposure of 1.3 millirems. (LIJ)

9/5/80 "Ertel's Suggestions Draw Fire."
A Congressional Task Force on TMI met. Cong. Ertel raised questions he thought the group could answer. Cong. Goodling

says some task force members have preconceived ideas and are bitter toward Met Ed. (HP)

9/5/80 "Cumberland Wants to be Part of TMI Impact Planning."
Cumberland County Commissioners criticized the NRC for exclusion of the county from a draft EIS. (HP)

9/6/80 "Met Ed Bankers Tighten Credit."
GPU's credit limit has been reduced by a consortium of 45 banks. (HP and LIJ)

9/9/80 "NRC Arguments Ripped by Panel."
A judges panel may rule that public hearings should have been held prior to the TMI venting. (HP)

9/10/80 "'No Way' Says PANE on Reopening of TMI Reactor 1."
The article describes PANE's plans to present legal and psychological arguments at the October hearings held by the NRC. (MPJ)

9/11/80 "Nuke Safety Expert Scores Maintenance Deferrals at TMI."
As a witness for the TMIA, R. Hubbard concluded that the owners of TMI practiced a policy of 'deferred maintanance' that made the plant unsafe and the 1979 accident worse. He was the safety consultant and technical advisor for the film *The China Syndrome*. (LIJ and HP)

9/12/80 "Squabble Could Delay Cleanup at TMI."
The article describes how the delay of the nomination of a new NRC Chairman could paralyze Unit I's restart and Unit II's decontamination. (HP)

"TMI Comment Extension Sought."
Gov. Thornburgh asks the NRC to double the time period allowed for public comment on the proposed TMI cleanup. (HP)

9/13/80 "Met Ed Plans to Cut 700 Jobs."
As part of an economy measure, Met Ed will eliminate 700 jobs, 500 of which are at TMI. (HP, LIJ, and NYT)

9/16/80 "TMI Sampling High in Tritium."
An analysis of soil 22 feet below the surface near Unit II reveals the highest level of radioactivity since the testing began. (HP)

9/17/80 "TMI Readings' Announcement Felt Unnecessary."
A GPU official says the company did not announce new high radioactivity levels because it didn't feel it was important for the public to know. (HP)

"March 28th Coalition Plans for Demonstration at TMI."
If a restart is ordered, the 3/28 coalition will demonstrate. (MPJ)

"Feds Begin Probe of TMI Bailout."
Cong. Ertel will head a congressional probe on the fiscal stability of Met Ed. (MPJ)

9/19/80 "Power Users Spared Cost of TMI Cleanup."
The PUC orders Met Ed to stop using customer revenues to cleanup the crippled contaminated N plant. The PUC also permitted Met Ed to pledge its accounts receivable to a consortium of banks in return for a $20 million boost in its credit ceiling. (LIJ and HP)

"GPU Lobbying for Bailout on TMI."
The financial crisis facing Met Ed causes GPU to lobby government and other N companies for help. (HP)

"TMI is Eyed as Political Issue."
Anti-N groups want to make TMI a presidential election issue. (HP)

9/23/80 "GPU Hesitates on PUC Rule."
A Met Ed attorney says the PUC order not to use consumer revenues in the TMI cleanup cannot be complied with without violating some federal or state laws. (HP and LIJ)

"3rd Entry of Building at TMI Set."
A third entry into Unit II is planned. Five men will make the longest trip to date. (HP; also 9/25/80 NYT)

9/24/80 "CNEC Sees Changes at TMI."
MCNEC tests TMI's security members, tour the facility, and learn of changes made since the accident. (MPJ)

"Denton and Stello are Lauded for Their High Performance."
Denton and Stello are honored with 47 other people by Pres. Carter. (MPJ)

"PUC Staff Calls for TMI Reactor Start Up."
The PUC staff urges the Commission to actively support the reopening of the undamaged Unit I N reactor at TMI. Met Ed desires to have the plant back in operation by 7/81. Arguments for and against the reopening are presented. (LIJ)

9/25/80 "NRC Says Control Room of TMI-1 is Inadequate."
A NRC inspection resulted in identifying 37 deficiencies in the Unit I control room. (HP; also 9/27/80 LIJ)

"TMI Building Entry Cancelled."
A PUC order prohibiting cleanup and recovery activities not covered by insurance has led to the cancellation of a planned manned entry at TMI. (HP)

9/26/80 "NRC Plans Thorough Check of Control Room at TMI-1."
The NRC says the identifying of 37 deficiencies at TMI I is just a

beginning. A more indepth probe is planned. (HP)

9/26/80 "Met Ed Wants Court to Break Statement." Met Ed will take legal action so that it can use customer revenues for some parts of the cleanup. It contends that the PUC order effectively prevents them from having the ability to cleanup TMI including a planned entry. PUC Shanaman denies Met Ed's charges saying they have sources of money other than rate payer money, such as insurance or shareholders. (LIJ)

9/27/80 "PUC is Upheld on Met Ed Rule." A Met Ed request to allow use on a temporary basis of customer funds for cleanup of TMI was rejected by US Middle District Court judge. (HP)

"NRC Insists its Orders on TMI are Paramount." The NRC says its concern over safety in decontamination should override the PUC ban on the use of consumer funds for the process. (HP)

10/1/80 "Met Ed Sues to Get Rate Hike." Met Ed has gone into court in pursuit of an emergency $35 million rate increase denied last month by the PUC. Met Ed states that its financial plight since the N accident causes the need for the increases. (LIJ; also 10/6/80 NYT)

"Orders are Restraining TMI Clean Up Activities." A TMI official charges that conflicting Pa. and federal orders are hampering the cleanup of TMI. (MPJ)

"Correction of Deficiencies at Unit 1 are Underway." TMI officials indicate that the problems in the Unit I control room are being corrected. (MPJ)

10/1/80 "Hearings on Unit Restart October 15th." ASLB initiates hearings on whether Met Ed should be permitted to restart Unit I. A list of witnesses is presented. (MPJ)

10/2/80 "TMI Issue 'Unpopular' in Congress." A TMIA spokesperson believes Congress is avoiding the issue of N cleanup at TMI because it is an unpopular subject in an election year. TMIA is demanding that Congress provide funding for public interest groups involved in litigation with the NRC. (LIJ)

10/3/80 "Audit Says GPU Faces Bankruptcy." "GPU is now a candidate to be the first major utility to go bankrupt since the depression," concludes Theodore Barry and Associates, a York consulting firm. The management audit was requested by the PUC and cost $775,000. The consulting firm's top priority recommendation was a joint task force to come up with a comprehensive plan of action after analyzing issues such as bankruptcy, reorganization, industry involvement, federal financial aid, federal takeover, and creation of a state power authority. (LIJ and HP; also 10/8/80 MPJ)

"Decontamination of TMI." A report on a *Science* magazine article states that TMI decontamination might be stalled indefinitely if no permanent disposal site is found for the radioactive wastes. (HP)

10/7/80 "NRC Says Vote on Issues at TMI Would Have Effect." Two hundred and fifty people attend a public meeting at F&M at which the NRC explains that the discharge of decontaminated water into the River would be safe. No decision has been reached. The NRC would consider results of a referendum if held. (LIJ)

10/9/80 "PUC-NRC Scrap Said Unplanned."
A PUC spokesman says that the PUC never intended to force a confrontation with the NRC over the TMI cleanup. (HP)

10/10/80 "Lessons Learned in TMI Crisis Soon Forgotten, Expert Claims."
In a speech before the American Institute of Mining, Metallurgical, and Petroleum Engineers, Harold Lewis said, "the forces of inertia make it very difficult for us to learn our lesson." Lewis is a physics professor and a member of the ACRS of the NRC. (LIJ)

10/11/80 "GPU View Mixed on NRC Control Room Ideas."
The NRC will propose control room changes if Unit I were ever to reopen. GPU's response is mixed. (HP)

10/12/80 "Public Relations Drive Aimed at Restarting Three Mile Island."
Met Ed takes reporters on a tour of the undamaged Unit I reactor at TMI. Restart hearings will begin on 10/15 before the ASLB. (MYT; also 10/15/80 MPJ)

10/14/80 "Building Entry Set at TMI."
The article details the planned third manned entry at TMI. (HP)

10/15/80 "PUC Chairman Named to US Nuclear Panel."
Susan Shanaman has been named to the Advisory Council of the National Association of Regulatory Utility Commissioners in D.C. The PUC also reported that N.E.I.L. will help pay utilities for the replacement power when accidents shutdown N plants. (LIJ)

10/16/80 "Safety of Nuclear Plants has Deteriorated Says NRC Engineer."
In letters to Congressmen and superiors at NRC, Demetrios L. Bedakas, a reactor safety engineer, contends that the safety of N plants has deteriorated since TMI.

All N plants should cut back to 65% of full power until 130 safety issues are resolved. (LIJ)

10/16/80 "N-Licensing Unit Opens TMI Hearings." The article reports on the hearings being held by the ASLB on Met Ed's request to return Unit I to operation. TMIA charges that the Unit should remain inoperative because the operator was incompetent and had neglected safety concerns in order to hold down operating costs. (MPJ and NYT)

10/17/80 "Members of TMI Hearing Panel Must Make The Tough Decisions."
The article describes Chairman Ivan Smith of the ASLB of the NRC. ASLB is conducting hearings into the restart of TMI Unit I. (LIJ)

"Entry at TMI by Team Called 'Very Successful.'"
A five man team enters Unit II building and performs the first maintenance since the accident. (HP; also 10/22/80 MPJ)

10/18/80 "NRC Creates 'Citizens Advisory Panel.'"
The NRC announces the creation of a 12 member citizen advisory panel. The panel will be consulted during the decontamination process. (HP and LIJ; also 10/22/80 MPJ)

10/20/80 "Anti-Nukes Write Book On What To Do If Unit I Reopens."
A decision concerning the reopening of Unit I is expected in several months. In preparation, TMIA has already written a 32 page training manual in methods of nonviolent protest and civil disobedience. (LIJ)

10/21/80 "Middletown Mayor Writes Carter for Federal Aid."
Mayor Reid appealed to Pres. Carter for federal funding for plant cleanup. He

alleged that if the area had been struck by a tornado or flood, the area would have received disaster help. He asks, Why not for TMI? (LIJ)

10/22/80 "No Need for US Aid in TMI Cleanup: SEC." A SEC analysis minimizing the need for federal aid at TMI drew skepticism and puzzlement from officials. Met Ed needs a one time $60 million federal loan guarantee assuming the undamaged Unit I reactor goes back on line by early 1981. (LIJ)

Editorial: "Federal Aid is Necessary to Make TMI Safe Again."
The editorial praises Mayor Reid's letter to Pres. Carter and urges federal aid is necessary to guarantee our health and safety. (LIJ)

"GPU Net Down 59%."
GPU, citing loss of TMI from rate base, says its net income in the 3rd quarter fell 59% to $10.5 million. (NYT)

10/23/80 "In Middletown, Election Issue is Economy Not TMI."
The economy is the key issue as derived from interviews with area residents. It is also noted that a group called "Friends and Families of TMI" was formed after the accident. (LIJ)

10/28/80 "Union Hits Met Ed Plan for Layoffs."
In a petition to the PUC, the IBEW says layoffs of 130 workers proposed by Met Ed will damage service in eight towns and cities. Met Ed has ten days to respond to the IBEW complaint. (LIJ)

10/29/80 "NRC Staffers Find TMI Charge Untrue."
A NRC investigation finds no evidence to support charges made by TMIA concerning lax maintenance and noncompliance with regulatory requirements at TMI. The report was made by The NRC's Office of

Inspection and Enforcement to the ASLB. (LIJ; also 11/5/80 MPJ)

10/29/80 "Met Ed Requesting $25 Million More." Met Ed has asked the PUC to grant a temporary $25 million surcharge on customer bills. The purpose is to improve the company's borrowing power with banks. (HP)

"Residents Discuss Cleanup with Officials." The article reports on a 10/20/80 PANE meeting where residents discussed cleanup with federal officials. (MPJ)

10/31/80 "Dumping TMI Waste in River May Affect Seafood Industry."
The NRC sponsored a meeting in Havre DeGrace, Md. focusing on plans to get rid of radioactive water. One hundred thirty residents, fishermen, and politicians attended the hearing and most of them feel that the dumping of treated water would have a disasterous effect on the struggling seafood industry. (LIJ)

"Long-Term Depression Prevalent in Mothers Living Near TMI Plant."
The results of a NIMH funded study indicate that 25% of the mothers assessed showed clinical levels of depression or anxiety during the year following the accident. (LIJ and NYT)

11/1/80 "TMI Reports Increase in Tritium Levels." An increase in Tritium is reported by TMI technicians, but levels are not considered dangerous. (LIJ; also 11/5/80 MPJ)

"Thyroid Birth Defects Up, Says M. Aamodt Blames Nukes."
M. Aamodt, a democratic candidate for the Pa. House, states that figures maintained by PDH supports the contention of a rise in thyroid birth defects which are TMI related. (LIJ)

11/3/80 "EPA Will Assess Clean Up at TMI."
The EPA will fund an independent scientific assessment of the cleanup options for the disabled TMI N plant. The group will examine the EIS released by the NRC. (LIJ)

11/5/80 "Unit 2 Decontamination Will Be Discussed."
A schedule of public hearings concerning Unit II decontamination is presented. (MPJ)

"Poll Reveals Public Confidence That Cleanup Will Be Safe."
A Field Research Corporation study for GPU indicates that a majority of the 2,849 people polled expressed confidence in the ability to cleanup TMI safety. (MPJ)

11/8/80 "TMI Cleanup Climbs to $1 Billion."
In comments to the NRC, a GPU official states that the cleanup may cost $1 billion. This is twice as much as earlier estimated. Also, the reactor core would not be removed until 8/85, two years later than anticipated. (LIJ and HP; also 11/12/80 MPJ)

11/10/80 "TMI Running Out of Sites for Disposal."
With the closing of the Hanford, Washington licensed disposal sites for low level wastes in 7/81, TMI is running out of disposal sites. Two other sites have not accepted wastes from TMI. Temporary crypts are being built at TMI. (LIJ)

"Nuclear Power Industry Examines Weaknesses."
The Institute of N Power Operations, which was set up as an arm of the industry after the TMI accident, plans recommendations to help operators fight boredom and the complacency that might occur when working around highly sophisticated machinery. (NYT)

11/11/80 Editorial: "N-Waste Peril."
The editorial urges Cong. Walker to assume a leadership role in getting the federal government to solve the problem of disposing N wastes. (LIJ)

"TMI Waste Storage Plan Only Temporary."
Arnold of Met Ed says its storage plan for TMI wastes is only temporary as the site is not an appropriate place for long term storage. (HP)

11/13/80 "Report Discounts Animal Mutations in Area of TMI."
A NRC report reviewed animal health problems on 22 farms. Complaints centered on animal mutation and vegetation stunting. It concluded that radiation was not the cause. (HP)

"TMI Panelists Aim Ire at NRC."
A citizen advisory committee on TMI cleanup met for the first time and criticized a federal report on the cleanup. (HP and LIJ)

11/14/80 "12 Take Extended Trip Inside Unit 2."
The article reports on the fourth and largest entry at Unit II to date. (HP, LIJ, and NYT)

11/15/80 "Videotape Shows Damage Inside TMI Containment."
The article reports on the videotape made during the fourth Unit II building entry. Among the findings are a melted telephone and a charred sign. (HP and NYT; also 11/19/80 MPJ)

11/19/80 "Final Cleanup Meeting."
The NRC scheduled a final public meeting concerning cleanup. (MPJ)

11/20/80 "NRC Chief Rapped at TMI Meet."
At the 31st and final hearing on testimony regarding an EIS, 100 people were present and one resident expressed most

of the audiences feelings when he said, "You don't listen to us. And we don't believe you." Chairman Ahearne responded, "Do we always listen to your comments? Absolutely. Do we always go with the majority? The answer is no." (LIJ and HP; also 11/26/80 MPJ)

11/20/80 "TMI Cleanup Cost Bill Mulled."
Pa. congressmen meet to discuss the type of TMI cleanup legislation to seek in Congress. (HP)

11/21/80 "GPU to Task Force Congressmen: Cash to Last at Least Till October."
GPU officials report to Pa. congressmen on revenue crisis. They expect the crisis will not occur before 10/81. (HP)

"Mayor Sees Stress if TMI Dumps Water."
In a letter to the NRC concerning an EIS, Lancaster Mayor Morris fears significant chronic mental stress in 20% to 40% of the population will occur if the decontaminated water is dumped into the Susquehanna River. (LIJ)

"TMI's Owners Say Funding Not Critical Now."
At a meeting of the Pa. Ad Hoc Task Force on TMI, executives of TMI stated that the company can operate without suffering a financial crisis for at least another year. (LIJ and HP)

11/22/80 "NRC Ignoring Lessons of TMI, Critic Claims."
In a hearing before the ASLB of the NRC, Dr. R. Pollard said, "the NRC is more interested in protecting the health and safety of the N industry than the public." (LIJ and HP; also 11/26/80 MPJ)

"NRC Says Animal Deaths Not Related to TMI."
A 38 page report concludes that the animal deaths were not related to TMI.

The study was done in collaboration with EPA and Pa. DOA. Janet Lee, an anti-N activist and farmer, disputes this conclusion. (LIJ)

11/26/80 "Nuclear Safety Report Criticizes Industry and Regulatory Agency."
The NSOC releases a letter to Pres. Carter that is critical of what it calls the 'business as usual mind-set' in the N industry and the NRC. (NYT; also 11/27/80 LIJ)

"'Bailing Out' TMI."
Several Pa. Congressmen are seeking $1 billion for the cleanup of the damaged TMI N reactor. (HP)

"State Utility Customers are Begging With PUC to Turn Down PEC Latest Rate Increase Request."
Met Ed and PEC customers beg the PUC to deny the companys' rate hike requests. The customers call for the federal government to assume responsibility in light of the TMI accident. (HP)

11/29/80 "US Could Get Stuck With Bill for TMI Cleanup."
In a draft report, the staff of the NRC say that the federal government could end up paying for cleanup of TMI. The cost is currently estimated at $900 million. Federal assumption could be avoided if the NJ and Pa. PUCs grant rate increases to GPU and Met Ed. (LIJ, HP, and NYT)

11/30/80 "Cost of 3 Mile Island Evacuation Estimated in Study at $9 Million."
A PSU study concludes that the evacuation after the accident at TMI cost individuals and business between $9 and $14.5 million. (NYT)

12/3/80 "Efforts to Stop Water Dumping Will Be Aired."
Testimony at a SVA meeting is highlighted including legal arguments of the PILC. (MPJ; also 12/4/80 HP)

12/4/80 "Restart TMI Unit I, GPU Head Urges Federal Agency."
GPU Pres. calls for immediate restart of Unit I. He accuses the NRC of discrimination in the delay. (HP and LIJ)

"Met Ed Seeks Quick OK to Restart Unit I."
Dieckamp, in a letter to NRC Chairman Ahearne, asks the NRC to expedite the restart of the undamaged reactor at TMI. Hearings began 10/15/80 by the ASLB and will continue through 3/81. Dieckamp asks for the restart of the reactor while the hearings continue. (LIJ and HP; also 12/10/80 MPJ)

12/6/80 "TMI Stress Factor Discarded by NRC."
The NRC ruled that the issue of psychological stress should be disregarded when its licensing board considers whether to restart the undamaged reactor at TMI. The vote was 2-2 constituting a denial of requests to admit stress as a factor. It will reconsider this issue when Pres.-elect Reagan appoints a fifth member to the NRC. PANE will appeal the decision. (LIJ and HP; also 12/10/80 MPJ)

12/8/80 Editorial: "No Short Cuts, Please, on Restarting Unit 1."
The editorial criticizes GPU's plan to startup the undamaged reactor while the ASLB continues its hearings. (LIJ)

12/9/80 "Reactor Owner Seeking $4 Billion From The US."
GPU files a $4 billion claim against the NRC contending the accident at TMI would not have occurred if the NRC had acted properly. GPU accuses the NRC of

failing to notify them of similar problems known as early as 9/77. (NYT; also 12/10/80 MPJ)

12/12/80 "Reactor Area at Three Mile Island Checked for High Radiation Level." A containment building of the damaged N reactor at TMI is explored and unexpectedly high radiation levels are found in the wall joints of the connecting building. (NYT, HP, and LIJ)

"Plant Coolant Test Succeeds." The DOE reports that Idaho National Engineering Laboratory simulated a N reactor accident like the one that happened at TMI and that the emergency core-cooling system worked successfully. (NYT)

12/17/80 "Federal Loan Guarantees Urged for TMI's Cleanup." Cong. Ertel outlined a plan to finance the $1 billion loan guarantees and some form of industry aid such as a DOE research grant to Met Ed. (LIJ and HP)

"Engineer Calls for Upgrade of Safety Systems at TMI." The testimony of Pollard of UCS before the ASLB is reviewed. (LIJ)

12/19/80 "DOE May Handle TMI's Reactor Core." In testimony to a TMI citizens advisory panel, Assistant Secretary Cunningham of DOE said it is quite reasonable to expect the DOE to assist in the treatment of the damaged, highly radioactive reactor core at TMI. In another development, intervenors in NRC restart hearings requested that the restart of TMI I should not be permitted until after the hearings have been completed. (LIJ and HP)

12/20/80 "The TMI Toll: Cleanup, Rates." A Pa. Governor's Office of Policy and Planning Report cited cleanup costs and higher electricity rates as the legacy of TMI. (HP)

12/24/80 "New Research Raises Questions on Danger of Nuclear Accidents."
The NSOC reports some preliminary evidence that radiological effects of N accidents may be substantially less than previously assumed. (NYT)

12/30/80 "TMI Stress Hardest on Educated."
Young, married, well-educated women tended to worry most about last year's N accident at TMI, a PDH report concludes. (LIJ)

12/31/80 "TMI Wants to Purify Water Now."
In testimony before a TMI citizens advisory panel, Arnold of Met Ed says they want to purify the radioactive water now even though it has not decided what to do with the water once it is treated. Treated water could be stored until 1982, on site, when a formal disposal plan will be prepared. Plant officials want to treat the radioactive water as it is a potential health hazard that could result in uncontrolled radiation releases. (LIJ and HP)

1/3/81 "$560 Million Suit."
Forty six Central Pennsylvanians filed a $560 million negligence suit against the owners, operators, and suppliers of TMI. (HP)

1/6/81 "Township Wants Met Ed to Pay Retroactive Fees."
The TMI plant was built when Londonderry Township charged $5 for building fees. Township supervisors want the company to pay a fair share under new construction rate schedule. (HP)

"TMI Plagued by Host of Problems but Radioactive Water Still No. 1."
A feature story concludes that something ought to be done here. But they do not

agree on what or how or when. In short, the cleanup of the crippled TMI II plant has come to virtual standstill. The story highlights interviews with government officials, public utility officials, and anti-N groups. (LIJ and NYT)

1/7/81 "Steam Release at TMI 'Routine.'"
A Met Ed spokesman says the recent steam release at TMI came from the nonradioactive portion of the idle Unit II system. (LIJ)

"GPU Pushes for Water Processing."
GPU seeks to divorce the issues of cleanup of waste water and the disposal of the same. GPU proposes use of the Epicor II system for cleanup while disposal talks proceed. (MPJ)

1/8/81 "US is Criticized on Lack of Plan in Reactor Crisis."
NSOC reports to Pres. Carter that the US has no policy on planning for emergency protection of 3.3 million people living near the country's 73 operating atomic power plants. It questions the NRC's readiness to intervene in the management of a stricken N reactor. (NYT)

1/13/81 "Top Court Backs SVA's TMI Cleanup Challenge."
The US Supreme Court refused to kill SVA's lawsuit challenging cleanup activities at the disabled TMI N plant. (HP and LIJ; also 1/21/81 MPJ)

Editorial: "A Victory for the SVA."
The editorial praises the SVA for persisting in this lawsuit. (LIJ)

"...While Jersey Central Claims Customers Should Pay for Idled Nuke Plant."
JCPL tells the Supreme Court that customers, not stockholders, should pay for Unit I cleanup. (HP)

1/14/81 "GPU's Financial Plan: A Fight for Survival."
GPU is fighting to avoid being the first utility since the depression to go bankrupt. Its plan includes suing B&W and the NRC; asking Pa. and NJ PUC for rate increases; asking other N utilities to help with cleanup; and asking for federal help. (LIJ and NYT)

"Brighter Financial Picture is Needed by PANE."
PANE has only a short time to decide if they will present an organized appeal to the NRC for consideration of psychological stress as a factor in TMI restart hearings. A big problem facing PANE is how to raise needed cash. (MPJ)

1/17/81 "Met Ed Rate Pitch 'Not Justified.'"
An expert witness for the Pa. consumer advocate testified that Met Ed's rate request is not justified. (LIJ)

1/20/81 "Tests Halted at TMI."
Inspection tests were halted after leaks were found. (HP and LIJ)

"PUC Puts Met Ed Rate Hike at $9.7 Million."
The PUC staff recommends a rate hike for Met Ed. (LIJ; also 1/21/81 MPJ and NYT)

1/21/81 "NRC Advisory Panel to Meet Here."
A public hearing by the NRC Advisory Panel on Decontamination of TMI will be held in Harrisburg. (MPJ)

"Explosives Found at TMI."
Deteriorated ammonia based crystals were found in a former concrete refuse area near Unit II. (MPJ)

"Spill of Sulphuric Acid at TMI."
About 500 gallons of sulphuric acid were spilled on the ground near Unit II. (MPJ)

1/21/81 "TMI Evacuation Blasted as Being 'Utterly Ridiculous.'"
The TMI PIRC has criticized federally funded studies on evacuation time estimates at nine N power plants including TMI. (MPJ)

1/22/81 "TMI-2 is now Cooling Itself."
Met Ed presented a public briefing on decontamination efforts. Unit II reactor is naturally cooling itself. The radiation recently found in test wells underground presents no threat to public health. (HP)

1/23/81 "TMI Slates 50% Cutback in Cleanup."
Cleanup cutback is ordered by Met Ed due to a PUC order barring use of customer revenue for uninsured cleanup costs. (LIJ)

1/28/81 "Update Given on Decontamination Efforts at TMI."
The article reports on a public hearing that was held in Harrisburg on TMI decontamination. (MPJ)

"Cork Sealant Contamination Found Not To Come From Unit 1."
Radioactive contamination found in the cork sealant joints of TMI Unit I is not coming from the reactor building. Studies indicate the source is the seal injection valve room in concrete walled cubicle of the auxiliary building. (MPJ)

"Met Ed May Sell Rotor Assembly."
An undamaged turbine rotor assembly from Unit II may be sold or leased to VEPCO. (HP)

1/29/81 "TMI Cited for Failure to Report Accident Data."
The NRC charges that Met Ed failed to immediately evaluate and report important data when the 1979 accident occurred at TMI. (HP, LIJ, and NYT)

2/4/81 "NRC Urges TMI Restart to Avoid Area Burnouts."
The NRC staff suggests a restart of the undamaged Unit I at power levels up to 5%. (LIJ; also 2/5/81 HP)

"Entry at TMI Paves Way for TV Installation."
Twenty two men installed equipment for closed circuit TV cameras inside the contaminated containment building. (LIJ and MPJ; also 2/6/81 HP)

"Test Successful at Unit I."
The article describes a series of tests at Unit I that were completed without problems. (MPJ)

"New Cooling Method at TMI Underway."
A new cooling method is in effect at Unit II. The unit is really cooling itself. The method is a losses-to-ambient-cooling and it replaces a steam-to-condenser method. (MPJ)

"NRC Staff Takes Enforcement Action Against TMI Utility."
The NRC has cited TMI for 2 items of alleged noncompliance. The items involve evaluation and reporting of information on the day of the accident. (MPJ)

"TMI Advisory Panel Will Meet Today."
The Citizens Advisory Panel to TMI will meet in Harrisburg. (MPJ)

2/7/81 "Ex-Congressman Tackles TMI Costs."
GPU hires ex-Cong. Thomas L. Ashley to develop a financial plan for the TMI cleanup. (HP; also 2/11/81 MPJ)

2/10/81 "Task Force Sees TMI Draft Bill."
A draft bill on TMI cleanup is being circulated among Pa. legislators. It calls for a national N accident insurance pool to pick up one fourth of the clean-up costs. (HP)

2/10/81 "TMI Committee Meeting to Study Rate Boycott."
A Newberry Township TMI Steering Committee meeting will discuss a rate boycott against Met Ed and information on non-violent civil disobedience in the event of restart of Unit I. (HP)

"TMI Chiefs Accused of Misleading State, NRC."
A report by the staff of the House Interior Committee concludes that the managers of the TMI N plant withheld information related to the severity of the reactor accident on 3/28/79, and made misleading statements to Pa. and federal officials. (LIJ and NYT)

2/11/81 "Cameras Rolling After TMI Entry Made."
The article reports on the sixth manned entry into Unit II. Seven of the eight cameras are working. (MPJ)

2/13/81 "Liability Costs of TMI System Irk Conoy Twp."
Two officials are upset that Conoy Township must pay for monitoring and wiring equipment. (LIJ)

2/14/81 "Panel Proposes Taxpayers Pay for TMI Cleanup."
The NOSC recommends that taxpayers pick up the estimated $1 billion tab to decontaminate the TMI N power plant. (HP and LIJ; also 2/15/81 NYT)

"Cleanup TMI, Ariz. Governor Says."
Gov. Babbitt says TMI should be cleaned up immediately and that wastes should be shipped to the DOE disposal site. He calls the matter a national and local concern. (HP)

2/17/81 "TMI Cleanup Plan Seen as Utility Customer Boom."
Pa. congressmen approve a proposal designed to secure federal help with

the TMI cleanup. Utility customers around the nation would pay as little as 37 cents annually. (HP)

2/18/81 "State Unit Gets TMI Cleanup Bill." The Pa. congressional delegation draft a bill that would fund most of the TMI cleanup and pay for future accidents at other reactor sites. (HP and LIJ)

"Rejection Urged of Plan to Merge TMI Management." An administrative law judge recommends that the PUC oppose a management merger proposed by 2 Pa. owners of TMI. (LIJ)

"PANE Accepts Appeal Challenge." PANE has filed an appeal to refute the NRC decision not to hear the issue of psychological stress in Unit I restart hearings. (MPJ)

"NRC Reduces $2,000 Fine Against Met Ed." Fines for violation of radioactive shipping regulations from TMI have been reduced by the NRC. (MPJ)

"Training Reported to Upgrade Quality at TMI." The article reports on the changes that have upgraded training for staff and management at TMI. (MPJ)

"Nationwide Electric Surcharge Asked to Pay for TMI Cleanup." A national electric surcharge idea is being proposed by Cong. Ertel. (LIJ)

2/20/81 "Using River Still One of TMI Options." Mayor Morris and Jean Kohr were the only advisory panel members who wanted the NRC to eliminate the use of the River as an option in its decontamination considerations. (LIJ)

2/21/81 "Morris Won't Give Up on TMI River Fight." Morris believes the advisory panel decision to use the River as a possible dumping site was a setback which will be corrected at a later date when the water waste disposal becomes a reality. (LIJ)

2/21/81 "Excessive Radioactivity Found Near Three Mile Island."
Radioactive Cesium levels in excess of federal drinking water limits were found in ground water near TMI. (NYT)

2/22/81 "Agreement Reached in 3 Mile Island Suit."
The owners and builders of TMI N plant reach a tentative $25 million settlement with thousands of claimants who live within 25 miles of the plant. (NYT)

2/24/81 "Nuke Insurance Bill OK Expected; Questions Raised on Task Force."
Reservations about the Pa. congressional task force bill on N cleanup are voiced on the eve of expected approval of the bill by the Pa. delegation. (HP)

"Unions Plan Protest for TMI Anniversary."
Eight international trade unions will pick up the anti-N banner and lead a protest marking the second anniversary of the TMI accident. (LIJ, HP, and MPJ; also 3/1/81 NYT)

"Nuclear Foe Reopens Charges of TMI Infant Death Rates."
Dr. Sternglass repeats his charges of TMI related infant deaths. (LIJ)

2/25/81 "Panel Asks Release of TMI Water."
A citizens advisory panel urges the NRC to approve a controversial demineralizer processing system to speed up the decontamination of radioactive water at TMI. (LIJ)

"Time to Address TMI Unit-1 Restart."
The ASLB will accept limited appearance statements on the restart of Unit I.

The hearing is set for 3/5 in Harrisburg. (LIJ)

2/28/81 "Radioactive Rats Pose Another TMI Problem."
Radioactive rodent droppings have been found in the basement of an uncontaminated service building at the TMI plant. (LIJ and NYT; also 3/4/81 MPJ)

3/3/81 "Ex-Nuclear Chairman Whom Carter Ousted is Named Acting Chairman."
Joseph M. Hendrie, who was dismissed by Pres. Carter as Chairman of the NRC amid criticism of its handling of the accident at TMI, is named Acting Chairman, replacing John F. Ahearne. The appoint- was made by Pres. Reagan and is viewed as an attempt to speed up the licensing process. (NYT)

3/4/81 "NRC Seeks Dismissal of PANE's Appeal."
The NRC has submitted a motion to dismiss PANE's appeal for consideration of psychological distress as an issue in the restart of Unit I. (MPJ)

"Citizens Plan Strike Against Met Ed Bills."
TMIA and Newberry Township TMI Steering Committee have called on Met Ed and PP&L customers not to pay their 3/81 electric bills. This is a form of protest against the cost of Met Ed electricity. The project is termed DAVID. (MPJ)

3/5/81 "Walker: Let Nuke Utilities Pick Up TMI's Cleanup Bill."
Cong. Walker supports Cong. Ertel's surcharge idea. He also favors the restart of TMI Unit I and the lifting of Met Ed's license to operate a N plant. (LIJ)

"N-Utilities are Silent on Cleanup."
The silence of electric companies on how their own property damage from N accidents should be financed perplexes the government. (HP)

3/6/81 "Unit 1 Restart Finds Scant Support at Hearing Here."
The article reports on a public hearing on the restart of Unit I. Most people urge that the unit not be reactivated. (HP and LIJ)

"Reagan OK's Aid for TMI Study."
Pres. Reagan has approved the inclusion of $27 million in the 1982 federal budget for a DOE project aimed at facilitating TMI cleanup. (HP and LIJ)

Editorial: "Found: The Magic World to Get TMI Cleaned Up."
The editorial lauds Pres. Reagan's decision. The editor urges the cleanup of TMI immediately and worry about the costs later. (LIJ)

"Need for Hearing on Venting Gas at Crippled Reactor is Upheld."
A federal appeals court upholds a ruling that the NRC should have held hearings on the release of radioactive gas from the damaged N plant at TMI. (NYT)

3/8/81 "Can Reagan Lift the Cloud Over Nuclear Power."
A review article assesses the status of the N energy industry two years after the TMI accident. (NYT)

3/10/81 "Cleanup May Take 7 Years at TMI; Risk 'Acceptable.'"
A NRC staff assessment about the cleanup is presented. (LIJ and NYT; also 3/11/81 MPJ)

"TMI Restart Hearing Off."
A special hearing on the restart of Unit I has been cancelled because William Penn Museum auditorium is not available. (HP)

"OK Ahead for Unit 2 Cleanup."
It is expected that the Reagan administration will make a substantial commitment to the TMI cleanup. (HP)

3/11/81 "Panel Asked to Draw Line on TMI Cleanup Spending."
The Reagan administration's funding for TMI cleanup has been questioned by Oregon's Sen. Hatfield. He believes the public will view the expenditure as a subsidy for GPU. (HP)

"PANE Says American Energy Week is Pro-nuclear Play."
American Energy Week is scheduled for 3/15-3/21. It is termed by PANE as an attempt to cloak the pro-N message in the mantel of an energy crisis. (MPJ)

"Request by Met Ed is Denied."
Met Ed's request that the courts reconsider their opinion that public hearings should have preceeded Krypton ventings is denied. The case is a result of PANE's objection to the venting. (MPJ)

"Residents Address Both Sides of TMI-I Restart Issue."
At a public NRC hearing on the restart of Unit I, residents spoke to both sides of the issue. Approximately 300 people attended. (MPJ)

3/13/81 "Bill Would Set Utility Insurance."
Cong. Ertel has introduced legislation that would force utility companies to establish a $750 million insurance fund to cleanup TMI and other N accidents. (LIJ and HP; also 3/18/81 MPJ).

"Conoy to Get TMI Siren Alert System From GPU."
GPU will install an emergency siren system, but not at the Township's expense. (LIJ)

3/14/81 "Unit 2 Entries Planned to Set Up a Monitor."
The article details plans for the seventh planned entry into the Unit II building. A remote monitor will be installed to measure radiation near the core. (HP; also 3/18/81 MPJ)

3/14/81 "Is TMI a Problem?"
A Dauphin County Commissioner tells the NRC that perceptions in D.C. are wrong regarding TMI. He speaks of the stalling and resident's anguish. (HP)

"Seminary Reaffirms Opposition to TMI."
The faculty of Lancaster Theological Seminary has reiterated its opposition to N power plants. (LIJ)

3/17/81 "Ex-Spaceman to Oversee TMI Cleanup."
Dr. James Fletcher, who headed NASA from 1971-77, will chair a scientific advisory board for GPU to make sure that the clean-up of TMI is done safely. (LIJ; also 3/18/81 MPJ and NYT)

3/18/81 "Udall to View Shutdown Reactors."
Congressmen Udall and Weaver will inspect Units I and II on 3/27. (HP)

3/19/81 "Big Pa. Tax Bill Faced by Met Ed."
The article describes how, without financial help, Met Ed will go bankrupt on 4/15 when $23 million in Pa. taxes fall due. (LIJ)

3/20/81 "Delayed Testing Begins at TMI."
The testing of filters to decontaminate radioactive water is initiated. (LIJ

3/21/81 "Met Ed's Rate Boost Shaved by Law Judge."
An administrative law judge recommends that Met Ed received two thirds or $58 million of a requested $76.5 million rate hike. (LIJ)

"Report Doubts Infant Death Rise From Three Mile Island Mishap."
PDH says radiation released in the accident at TMI N plant on 3/79 did not cause the deaths of infants or fetuses. George Tokuhata, Director of Pa. BER, comments. (NYT)

3/24/81 "NRC Rejects Restart of TMI Unit 1."
The NRC refuses to allow a restart of Unit I until the completion of hearings into the N plant's condition. (HP and LIJ; also 3/25/81 MPJ)

3/25/81 "March Rally This Saturday for TMI Anniversary."
The article details plans for the second anniversary march on TMI. (MPJ)

"M-Town Association Authors Resolution for Unit I Restart."
The MAA of the CC of the Greater Harrisburg Area has passed a resolution in support of reopening Unit I. (MPJ)

"'Footprints' of TMI Accident Linger On."
Unanticipated contaminations continue to appear at TMI. It is believed that these problems will remain until the radioactive water is removed. (MPJ)

"NRC Advisory Group Meeting on TMI."
The NRC will hold a technical meeting in D.C. to review modification made to the Unit I plant. (MPJ)

"The Cleanup-GPU Draws All Forces."
The article reviews progress on TMI cleanup, updates schedule, and presents GPU a view of the process. (MPJ)

"TMI is His Path in Daily Jogging Duty."
A photo story of a TMI employee who jogs around the cooling tower next to the Unit II reactor during lunch hours. (MPJ)

"GPU Maintains SDS System is Cure."
The story describes the differences between the Epicor II System and the SDS System for removing waste water at TMI. GPU is said to favor the SDS. (MPJ)

3/26/81 "TMI Cover-Up Charged."
Dr. Ernest Sternglass accuses PDH of manipulating infant mortality figures

after the TMI accident as part of a conspiracy to save the industry. (HP)

3/26/81 "Idled TMI Unit I Clouds Lebanon Business Future."
A report says the business future in Lebanon is cloudy due to the increasing costs of electricity which are related to the inactivity of Unit I. (HP)

"Professor's Accusations Rebutted."
The PDH admitted its 1979 vital statistics report was late but it denies the delay was used to doctor figures relating to TMI. (HP)

"Restart of Unit I Urged by Chamber."
The Pa. CC says idled Unit I has hiked utility bills and weakened Pa's business climate. It urges restart of the unit. (HP)

"How TMI Water Would be Cleaned."
The article describes how demineralizer processing system will operate. (LIJ)

"Anti TMI Rally Gains Labor Support."
The supporters of an anti-TMI rally are identified. (LIJ)

"TMI: 2 Years Later."
This summary article indicates that the massive cleanup is now 7 years away. (LIJ)

3/28/81 "Fired-Up at Met Ed, 200 Anti-Nukes Burn $47,000 in Bills at Capitol Steps."
On the eve of the second anniversary of TMI, nearly 200 anti-N protestors fed utility bills to a bonfire on the steps of the State Capitol. (HP and LIJ)

"NRC Cleanup Funds are Opposed in D.C."
The House Interior Committee has prohibited the NRC from spending federal funds to cleanup Unit II. Similar action is expected later in relation to DOE funding. (HP)

3/28/81 "End of Negligence-Claim Ban Spurs 24 TMI Lawsuits."
As the statute of limitations ran out on alleging negligence in connection with the TMI accident, 24 law suits are filed. (HP)

"Unit 1 Restart Gets Support from Udall."
After inspecting TMI, Cong. Udall said Unit I should probably be allowed to restart. (HP and NYT)

Editorial: "Two Years Ago Today."
The editorial argues that the primary lesson of the TMI accident is that N power plants should never be built in major population centers. (LIJ)

"Suit Filed for State Expenses From TMI Mishap."
Attorney General Leroy Zimmerman files a federal lawsuit to recover more than $10,000 which he said is the extra operational and overtime expenses Pa. incurred by the TMI N accident. (LIJ)

FEDERAL AND STATE GOVERNMENT DOCUMENTS

In this section, we present annotations of the NRC documents, the Kemeny Commission Reports, and other federal and state government documents concerning TMI. These annotations will convey to the reader the issues that were major concerns of various government bodies.

In reading these annotations, two problems are identified. First, the difficulty of keeping up with the numerous government publications that are published. These publications are released by many sources such as the NRC, EPA, special commissions and congressional hearings.

Another problem is the use by the NRC of corporations with vested interests to evaluate major issues concerning the nuclear power industry. It is questionable whether the reports of corporations like Bechtel Power can be unbiased. At the same time, it must be recognized that very few independent organizations exist that have the objective capabilities required by the government.

Because these government documents are not easily accessible, they are annotated for the reader. We hope that our non-value annotations will help the reader to be informed about the government's role in the key issue of commercial nuclear power.

A. NRC Documents

NRC documents relating to TMI in particular and some to nuclear power in general are annotated below. The documents are presented in chronological order employing the numbering system of the NRC. These documents can be read at government libraries (see back for more information) or can be purchased from the NRC. For more information about purchasing, write USNRC, Washington DC 20555 and/or National Technical Information Service, Springfield, VA 22161.

NUREG CP-0009; *Conference on Radiological Emergency Training Held 6/24-25, 1979*

As a result of accident at TMI, program for improving State and local radiological emergency response plans are receiving substantial attention. Review of plans were made in light of NRC criteria.

NUREG CP-0011; *Proceedings of Workshops on Proposed Rule Making on Emergency Planning for Nuclear Power Plants 4/80*

Transcripts of four workshops held in 1/80 in different areas of U.S. for NRC to receive comments concerning proposed rule changes.

NUREG-0020; *Operating Units Status Reports; Vol. 1-Vol.5*

Monthly reports began in 1977. Each report contains monthly highlights and statistics for commercial operating nuclear power units.

NUREG-0030; *Construction Status Report of Nuclear Power Plants*

Quarterly reports began in 1978. Report includes data from utilities sponsoring the construction, as well as from NRC's Office of Inspection and Enforcement. Data is used to synchronize the licensing process with predicted fuel load dates.

NUREG-0040; *Licensee Contractor and Vendor Inspection Status Report*

Vol. 1 began in 1977. Quarterly reports list each power plant inspected and a summary of the findings are noted.

NUREG-0090; *Report to Congress on Abnormal Occurrences*

Quarterly reports of commercial operations starting with Vol. 1 in 1978.

NUREG-0107 Supplement #2; *Safety Evaluation Report Related to the Operation of TMI-2 2/78*

The general evaluation of TMI II was unchanged indicating that it was a safe plant to operate. One of the issues resolved in this evaluation was the approval of emergency cooling analysis modifications.

NUREG-0348; *Demographic Statistics Pertaining to Nuclear Power Reactor Sites 10/79*

The siting of nuclear power reactors continues to be a subject of NRC interest. It is the intent of this report to provide a consolidation of data which deals with the population in the areas surrounding nuclear power reactors, so as to permit evaluations of the aggregate results, including trends of siting practices. The report provides population statistics for 145 nuclear power plant sites.

NUREG-0380; *Program Summary Report: Vol. 1-Vol. 5*

Each report, starting in 1977, summarizes the status of nuclear facilities in the U.S. including reactor licensing, construction and operation, standards and regulations, nuclear materials, and research projects. Reference to TMI is made in each report.

NUREG-0454; *Effects of Nuclear Power Plants on Community Growth and Residential Property Values 4/79*

The study tested the hypothesis that nuclear power plants adversely affect community growth and residential properties in near municipalities. The primary finding was that the plants exerted no influence on the price of housing. Therefore, the original hypothesis is rejected.

NUREG-0535; *Review and Assessment of Package Requirements (Yellow Cake) and Emergency Response to Transportation Accidents; Final Report 7/80*

Following the accidental spill of yellow cake in Colorado in 9/77, a USNRC/US Department of Transportation Study Group met to consider topics on yellow cake packaging and response to transportation accidents involving radioactive material.

Yellow cake is an uranium ore concentrate. The group concluded that in an accident the state and local government agencies are responsible for controlling the scene, that carriers are responsible for notifying authorities and isolating and cleaning up any spilled radioactive material, and that the shippers are responsible for informing others of hazards in the cargo. The group recommended that these parties prepare plans for carrying out these responsibilities. The group also recommended transportation data gathering programs, but, based on cost effectiveness arguments, did not recommend additional package requirements for yellow cake shipments.

NUREG-0540; *Title List of Documents Made Publicity Available by NRC*

Vol. 1 #1 is 1/79; Vol. 2 #1 is for 1/80; and Vol. 3 #1 is for 1/81. Monthly listing is presented by nuclear power plants including TMI I and TMI II.

NUREG-0555; *Environmental Standard Review Plans for the Environmental Review of Construction Permit Applications for Nuclear Power Plants 5/79*

The ESRP is prepared for the guidance of staff reviewers in the ONRR in performing reviews of each construction permit application.

NUREG-0568; *Title List Publicly Available Documents TMI Unit 2 Docket 50-230 5/79*

Cumulated listing of preaccident documents and after accident to 5/21/79, Rev. 1, Supp. 1 7/1/79-10/31/79 and NUREG-0631 has the cumulated listings until 11/16/79.

NUREG/CR-0570; *Technology, Safety and Costs of Decommissioning a Reference Low Level Waste Burial Ground 6/80*

Vol. 1 is the main report and Vol. 2 contains the appendicies.

The purpose of this study is to provide information on the available technology. The safety considerations and the probable costs of decommissioning low level waste burial grounds after waste emplacement operations are presented. This information is intended for use as background data in the development of regulations pertaining to decommissioning activities. It is also intended for use by regulatory agencies and site operations in developing improved waste burials and site maintenance procedures at operating burial grounds.

NUREG-0585; *TMI-2 Lessons Learned Task Force Final Report* 10/79

The primary conclusion of the task force is that, although the accident at TMI stemmed from many sources, the most important lessons learned fall in the general area called operational safety. Specifically, inadequate attention has been paid by all levels and all segments of the technology to the human element and its fundamental role in both the prevention of accidents and the response to accidents.

NUREG-0591; *Environmental Assessment: Use of Epicor II at TMI, Unit 2* 8/14/79

This study concluded that the proposed use of Epicor II for the processing of contaminated waste from the TMI Unit II auxiliary building will not significantly affect the quality of the human environment. Therefore, the Commission has determined that an EIS need not be prepared and that, pursuant to 10 CFR 51.5 (c), issuance of negative declaration to this effect is appropriate.

NUREG-0569; *Non-Radiological Consequences to the Auqatic Biota and Fisheries of the Susquehanna River from the 1979 Accident of TMI Nuclear Station* 11/79

Non-radiological consequences were assessed through the post-accident period of 7/79. Thermal and chemical discharges during the period did not exceed required effluent limitations. No impacts on benthic invertebrates or fishes were detected.

NUREG-0610; *Draft Emergency Action Level Guidelines for Nuclear Power Plants* 9/79

This document is provided for interim use during the initial phases of the NRC effort to promptly improve emergency preparedness at operating nuclear power plants. Public comment is being solicited and the deadline was 12/1/79.

NUREG-0618; *Nuclear Power Plant Operating Experience, 1978 Annual Report*

The fifth annual report summarizes the operating experience of US nuclear power plants in commercial operation. Power generation statistics, plant outages, reportable occurrences, fuel element performance, and occupational radiation exposure for each plant are discussed.

NUREG-0625; *Report of the Siting Policy Task Force 8/79*

The Task Force was established by the NRC in 8/78 to develop a general policy statement on nuclear power reactor siting. The report was completed before the TMI accident. A recommendation of the report is to include population density as a criterion in siting.

NUREG-0628; *NRC Staff Preliminary Analysis of Public Comments on Advance Notice of Proposed Rulemaking on Emergency Planning 1/80*

Proposed rules are printed along with public comments.

NUREG-0632; *NRC Views and Analysis of the Recommendations of the President's Commission on the Accident at TMI 11/79*

The report analyses and comments on recommendations in relation to the structure of the NRC, the utility and its supplies, training of operating personnel, technical assessment, worker and public health and safety, emergency planning and response, and the public right to information.

NUREG-0636; *The Public Whole Body Counting Program Following the TMI Accident 12/80*

Whole body counter is a device which has the capability of measuring very small quantities of radioactivity in people. This method was used to assess 753 men, women, and children. No radioactivity was identified by this method which could have originated from the radioactive materials released following the accident at TMI. Technical problems encountered, results, and conclusions are discussed.

NUREG-0637; *Report to the NRC from the Staff Panel on the Commission's Determination of an Extraordinary Nuclear Occurrence 1/80*

The report concludes that the accident at TMI did not result in offsite radiation discharge or offsite damage to property or people. Therefore, the accident at TMI did not constitute an "extraordinary nuclear occurrence."

NUREG-0640; *TMI, Unit 2, Radiation Protection Program; Report of the Special Panel Office of Nuclear Reactor Regulation 12/79*

1. The Panel confirmed several management and technical deficiencies in the program. However, recent major GPU/Met Ed commitments and actions demonstrated a major positive change in management attitude.
2. The Panel concluded that radiation exposures to personnel can be maintained low as is reasonably achievable while limited preparatory recovery work continues and when further needed improvements are implemented as needed, the radiation safety program will be able to support major recovery activities.

NUREG-0642; *A Review of NRC Regulatory Processes and Functions 1/80*

A review of the history of NRC regulatory functions is presented with an analysis of weaknesses with recommendations for correction of these weaknesses in light of the accident at TMI.

NUREG-0648; *Study of the NRC's Appellate System 1980*

The study examines: (1) the history of appeals board, (2) the role the appeal boards play in the current adjudicatory process, (3) legislative trends relating to use of intermediate appellate bodies for agencies, (4) the adjudicatory system at other agencies, and (5) the current workload of the appeals board and the commission. The study concludes with a discussion of available options and makes recommendations for potential improvements.

NUREG-0654; FEMA Report 1. *Criteria for Preparation and Evaluation of Radiological Emergency Response Plans and Preparedness in Support of Nuclear Power Plants; For Interim Use and Comment 1/80*

NUREG-0660; *NRC Action Plan Developed as a Result of the TMI-2 Accident: Vol. I and II 5/80*

In Volume 1, there is presentation of NRC's plan in relation to operational safety, siting, design, emergency preparedness and radiation effects, practices and procedures and policy, and reorganization and management. In Volume 2 all the tables developed by the five task forces are presented.

In the area of emergency preparedness, it was concluded: "that in light of the TMI accident, the overall state of planning and preparedness for nuclear emergencies was inadequate."

NUREG-0662; *Final Environmental Assessment for Decontamination of the TMI Unit 2 Reactor Building Atmosphere: Vol. 1 and 2 5/80*

Recommendations:

1. The potential physical health impact on the public of using any of the proposed strategies for getting rid of the Krypton 85 is negligible.
2. The potential psychological impact is likely to grow the longer it takes to reach a decision, get started, and complete the process.
3. The purging method is the quickest and the safest for the workers in TMI to accomplish.
4. Overall, no significant environmental impact would result from use of any of the alternatives discussed in this Assessment.

NUREG-0662; *TMI Legal Fund Comment 4/14/80*

The NRC proposal to vent Krypton 85 is "replete with errors of both fact and judgment." The reasons for this conclusion is as follows:

1. No emergency at hand;
2. Proposed venting "carries definite genetic and carcinagenic risks to the people of nearby communities;" and

3. Proposed venting "cannot be controlled due to meteorologic uncertainty."

NUREG/CR-0671; *Decommissioning Commercial Nuclear Facilities: A Review and Analysis of Current Regulations 8/79*

This report describes and analyses the regulatory requirements and guidelines applicable to the decommissioning of commercial light water reactors, other commercial nuclear fuel cycle facilities, and by product utilization facilities, as contained principally in the US code of Federal Regulations and the US NRC's Regulatory Guides. State requirements are discussed where appropriate. The report provides general background information to license applicants and to other interested parties. Included is an outline of procedural steps required of an applicant to comply with decommissioning regulatory requirements.

NUREG/CR-0672; *Technology, Safety, and Costs of Decommissioning a Reference Boiling Water Reactor Power Station*

Vol. I is the main report and Vol. 2 contains the appendicies.

1. Three approaches to decommissioning: immediate dismantlement, entombment, and passive safe storage with deferred dismantlement are studied to obtain comparisons between costs, occupational radiation doses, potential radiation dose to the public, and other safety impacts.
2. The primary conclusion is that the safety impacts of the decommissioning operations on the public are found to be small with the principal impact on the public being the radiation dose resulting from the transport of radioactive materials to a disposal site.

NUREG-0673; *Answers to Questions About Removing Krypton from TMI-2, Reactor Building*

The purpose of this staff report is to provide a clear, readily understandable description of the alternative proposals for removing the Krypton and the possible impact of each alternative. Questions and answers approach is utilized.

NUREG-0680; *TMI-Restart, Evaluation of Licensee's Compliance with the Short and Long Term Items of Section II of the NRC Order Dated 8/9/79*, Met Ed Co. *et al 11/80*

"Based on our review to date, we conclude that for a significant majority of individual items and sub-items of the order, the licensee is in compliance with the requirements of the order. Discussions with the licensee are continuing on the remaining open items, and after submittal of additional required information and completion of our review, we will report further on these matters in a supplement to this evaluation.

"Any recommendations by the staff to authorize restart of TMI I will be made only after further resolution of presently open items, definition of any additional required items, and compliance by the licensee with those additional requirements."

NUREG-0681; *Environmental Assessment of Radiological Effluents From Data Gathering and Maintenance Operation on TMI Unit 2 5/80*

The conclusion was that the environmental impact associated with this action will not exceed those already described in the FES of 1972 and 1976 for TMI. Therefore, no significant environmental impact will result. (Interim Criteria Approved by the NRC on 4/7/80)

NUREG-0683; *Draft Programmatic EIS related to the Decontamination and Disposal of Radioactive Wastes Resulting from 3/28/79 Accident TMI Nuclear Statics, Unit 2 7/80*

The staff has found that methods exist or can be suitably modified to perform all these operations with minimal releases of radioactivity to the environment.

NUREG-0684; *Summary of Public Comments and NRC Staff Analysis is Relating to Rulemaking on Emergency Planning for Nuclear Power Plants-USNRC 9/80*

Assessment of public comments is presented with a publication of the final regulations adopted concerning emergency planning.

NUREG-0689; *Potential Impact of Licensee Default on Cleanup of TMI-2 11/80*

The financial repercussions of the accident at TMI II on the ability of Met Ed to complete the cleanup of the facility is examined. "Potential impact of licensee default on cleanup and alternatives to minimize the potential of bankruptcy are discussed. Specific recommendations are made regarding steps the NRC might take in keeping with its regulatory functions and its mission to protect the public health and safety."

NUREG-0694; *TMI Related Requirements for New Operating Licenses 6/80*

This report summarizes the several parts of the list of TMI related requirements approved by the NRC for new operating licenses.

NUREG-0698; *NRC Plan for Cleanup Operations at TMI Unit 2*

This plan defines the NRC's role in the cleanup operations at TMI II and outlines the NRC's regulatory responsibilities in fulfilling this role. These responsibilities include reviewing and approving Met Ed's proposals for cleanup actions, overseeing the licensee's implementation of approved activities, coordinating with other federal and state governmental agencies on their activities in the cleanup and informing local officials and the public about the stations of the cleanup operations.

NUREG-0728; *Report to Congress: NRC Incident Response Plan 9/80*

NUREG-0729; *Report to Congress on NRC Emergency Communications 9/80*

NUREG-0730; *Report to Congress on the Acquisition of Reactor Data for the NRC Operations Center 9/80*

Collectively, these reports provide a comprehensive outline of the actions and plans for the NRC for improving its response to any future accidents.

NUREG-0732; *Answers to Frequently Asked Questions About Cleanup Activities at TMI-Unit 2* 9/80

Question and answer approach used to discuss purpose, scope, public protection, and all aspects of the decontamination process. The staff concluded that, on balance, "the benefits of fuel decontamination, fuel removal, and disposal of the radioactive wastes greatly outweigh the environmental costs of the cleanup."

NUREG-0737; *Clarification of TMI Action Plan Requirements* 11/80

This document is a letter from D.G. Eisenhut, Director of the Division of Licensing, NRC, to licensees of operating power reactors and applicants for operating licenses forwarding post-TMI requirements which have been approved for implementation. It is based on the action plan encompassed in NUREG-0660.

NUREG-0738; *Investigations of Reported Plant and Animal Health Effects in the TMI Area* 10/80

This report concludes that none of the reported problems could be associated with the accident at TMI.

NUREG/CR-0744; *Identification and Assessment of the Social Impacts of Transportation of Radioactive Materials in Urban Environments-NRC* 7/80

Based on a review of incidents involved in transportation accidents, the study concluded that the "political and legal impacts have been more substantial than psychological and social impacts, and for a given magnitude of physical consequences, radioactive transportation has greater social impacts than hazardous materials transportation generally."

NUREG/CR-0745; *Measured and Predicted Gas Flow Through Rough Capillaries* 6/79

The purpose of this analysis is "to help predict the flow of airborne particles through microscopic leads in plutonium dioxide shipping containers during postulated accidents, a comparative study of experimental and calculated gas flow rates through rough metal microcapillaries was carried out." The computer model (CAPIL) was used and the specific technical conclusions are presented.

NUREG-0746; *Emergency Preparedness Evaluation for TMI 1 12/80*

The TMI I Emergency Plan generally meets the requirements except for several specific items which are identified. The licensee must conduct an emergency preparedness exercise with the state and county governments to show that the plan can be implemented satisfactorily. A finding on the state of preparedness in the environs around the site is due from the FEMA. A supplement to this report, will include this finding and will address the other items that currently need resolution.

NUREG/CR-0774; *Early Mortality Estimates for Different Nuclear Accidents. Final Phase I Report (10/77-4/79) 8/79*

The purpose of this report was to analyze existing data as a basis for further research. A computer-based simulation model was developed which predicts early mortality in populations exposed under known conditions. "The end result is an estimate of frequency of effects as a function of the doses to critical organs."

NUREG/CR-0849; *Safeguards Material Control and Accounting Program: Quarterly Report*

Activity for each quarter (Vol. 1 was started in 12/78) in the Material Control Safeguards Evaluation Program, conducted for the USNRC at Lawrence Livermore National Laboratory, is summarized.

NUREG/CR-0912; *Geoscience Data Base Handbook for Modeling a Nuclear Waste Repository Vol. I and II 1/81*

"These handbooks contain representative data useful for assessing radiological risks associated with repositories for high level, transuranic, and spent-fuel wastes. The data will also aid in the derivation of deep geological modeling parameters and will serve as standards for weighing other data presented in licensing applications for deep geological waste repositions."

NUREG/CR-0913; *Generation of Hydrogen During the First Three Hours of the TMI Accident* 7/79

"Estimates of the amount of hydrogen generated by oxidation of fuel cladding during the first three hours of the TMI accident are presented. Fuel cladding is the first of three barriers isolating fission products from the environment. Our best estimate is that about 35% of the total core fuel cladding was oxidized producing 350 KG of hydrogen by three hours after scram."

NUREG/CR-0958; *Analysis of Particulate Transmission Through Small Openings Resulting From Container Stresses* 4/80

"This report concludes a study relating particle transmission to gas leakage through micro-opening in containers subjected to stress." An airplane crash was used as "stress." The results indicated that "the loss of an amount of plutonium oxide exceeding the allowable standards (interpreted as 2550 ug) is extremely unlikely."

NUREG/CR-0964; *A Feasibility Study for a Computerized Emergency Preparedness Simulation Facility* 11/79

This report details the feasibility of a computerized Emergency Preparedness Simulation Facility for use by the NRC. This approach is being studied as the current "drills often do not provide enough realism to actual emergency conditions."

NUREG/CR-0970; *Licensee Performance Evaluation: Phase II* 8/79

"It was found that licensee event report indicators could be more easily identified and utilized than noncompliance indicators based on presently available data systems." A model is being proposed as having the advantage of avoiding individual inspector and/or regional variations in evaluation.

NUREG/CR-0990; *The Use of Reconnaissance Level Information for Environmental Assessment* 11/79

RLI is "sufficient for comparing the environmental and socio-economic features of candidate sites for nuclear power stations and for guiding plant design, baseline surveys, and operational practices. It is usually available from published reports, public records, and knowledgeable individuals. Environmental concerns for site evaluations include: aquatic ecology, terrestrian ecology, land and water use, socio-economics, and institutional constraints." A scheme for using RLI to assign classification for candidate sites is proposed.

NUREG/CR-1018; *Review of LWR Fuel System Mechanical Response with Recommendations for Component Acceptance Criteria* 9/79

"Earthquakes and postulated pipe breaks in a nuclear reactor coolant system would result in external forces on the reactor fuel system. The effects of these applied forces on fuel system coolability must be assessed to insure reactor safety and hence a fuel assembly acceptance criteria must be developed to allow for this assessment." The rest of the report deals with the establishment of these criteria.

NUREG/CR-1060; *Activities, Effects and Impacts of the Coal Fuel Cycle for a 1,000-MWE Electric Power Generating Plant Final Report* 2/80

The report was prepared so that it could be compared with the effects and impacts of the uranium fuel cycle for 1,000-MWE nuclear power plant.

NUREG/CR-1093; *TMI Telephone Survey: Preliminary Report on Procedures and Findings* 10/79

The data from this study indicate that the accident at TMI affected a large number of people, both socially and economically, and that it is continuing to affect some people. Thirty nine percent or 144,000 people perceived the danger, the confusion, and the fear of a forced evacuation. See NUREG/1215 for Final Report

NUREG/CR-1125; *Population Dose Commitments Due to Radioactive Releases from Nuclear Power Plant Sites in 1976* 12/79

"Population radiation dose commitments have been estimated from reported radio-nuclide releases from commercial power reactors during 1976. Fifty year dose commitments from a one year exposure were calculated from both liquid and atmospheric releases for four population groups (infant, child, teenager, and adult) residing between 2 to 80 KM from each site. The data for TMI is included."

NUREG/CR-1131; *Examination of Offsite Radiological Emergency Protective Measures for Nuclear Reactor Accidents Involving Core Melt*

Originally published in 6/78 and reissued in 10/79 in response to renewed interest generated by the accident at TMI.

"Evacuation, sheltering followed by population relocation and iodine prophylaxis are evaluated as offsite public protective measures in response to nuclear reactor accidents involving core melt." Recommendations are provided for different types of core meltdown including consideration of atmosphere conditions.

NUREG/CR-1152; *Recommended Methods for Estimating Atmospheric Concentrations of Hazardous Vapors After Accidental Release Near Nuclear Reactor Sites 4/80*

This literature survey of atmospheric transport and dispersion characteristics of hazardous substances provides a basis for recommendations for input to the evaluation of postulated accidents in the vicinity of nuclear reactor sites.

NUREG/CR-1205; *Data Summarizes of Licensee Events Reports of Pumps at U.S. Commercial Nuclear Power Plants 1/80*

This report describes the results of an analysis of nuclear plant pump failures. The data used for this analysis were the LERs. LERs are reports filed with the NRC by operators of plants. Information for TMI is included for the years 1/1/71-4/30/78.

NUREG-1215; *The Social and Economic Effects of the Accident at TMI-Findings 1/80*

A. Economic Effects
Interrupted local production and reduced income and employment during the first week of April, but minor since then. Total short term losses about $9 million.

B. Institutional Effects
1. No formal emergency was declared. The role of C.D. coordinators was ambiguous.
2. Lack of a specific evacuation plan prior to the accident complicated the work of local emergency agencies.

C. Individual Effects
1. About 1/3 of the population of 370,000 within 15 miles of the plant evacuated.
2. For most people the effects of the accident were short-termed. "The extent of the continuing anxiety will depend on their participation in the decision-making process, on their ability to understand the logic of the decisions that are made, and the credibility of the decision-making bodies."

NUREG/CR-1219; *Analysis of the TMI Accident and Alternative Sequences 1/80*

Utilizing the MARCH computer code, an analysis of the technical aspects of why the accident occurred at TMI is made. Furthermore, alternative possibilities are also explored.

NUREG-1250; *TMI: A Report to the Commissioners and to the Public 1/80*

Vol. 1 contains a narrative of the accident, a summary, and the recommendations of the Rogovin Panel. Part I of Vol. 2 focuses on the pre-accident licensing and regulatory background. This part includes an examination of the over-all licensing and regulatory system for N power plants viewed from different perspectives including an analysis of TMI. The primary conclusion is that the "action taken to enable completion of TMI II in 1978 did not comprise the safety of the unit."

NUREG-1250; Vol. 2, Part II 1/80

1. The direct social and economic effects of the TMI II accident were dramatic in terms of short term disruption, but were mostly transitory.
2. The accident's most significant effect on the people was the evacuation experience. In a climate of confusing and conflicting information, pressures to evacuate mounted within the population from the 1st day, 3/28/79. By the time of the Governor's advisory on Friday at 12:30 PM, 14% of those who would evacuate had already done so. Within 15 miles, an estimated 144,000 people evacuated (39% of the population).

Recommendations

1. Improved public education
2. Improved information flow

This part focuses on a technical description of the accident. It includes a narrative description of the accident, a time line chronology, a discussion of radioactive releases and the radiation protection program at TMI, an assessment of plant behavior, a discussion of core damages and alternative accident scenearios, and a discussion of human factors.

NUREG-1250; Vol. 2, Part III 1/80

This part focuses on the response to the accident, safety management factors germane to the accident, a list of dispositions and the findings, and recommendations of Vol. 2.

NUREG/CR-1270; *Human Factors Evaluation of Control Room Design and Operation Performance at TMI-2 Vols. I-III 1/80*

Report for the TMI Special Inquiry Group. An analysis of the first 150 minutes of the accident was conducted to assess control room design problems and human errors involved in the accident at TMI.

NUREG/CR-1280; *Power Plant Staffing 3/80*

"This report discusses the results of a comparative review of the selection, training, qualification, and requalification of 1) maintenance personnel 2) operators 3) shift supervisors and 4) senior onsite managers involved in the operation and/or maintenance of N power plants. It also contains recommendations to improve the NRC requirements and civilian practices."

NUREG/CR-1331; *Data Summaries of Licensee Event Reports of Control Rods and Drive Mechanisms at US Commercial Nuclear Power Plants 1/1/72-4/30/78 2/80*

Same procedure as described in 1205.

NUREG/CR-1353; *Preliminary Calculations Related to the Accident at TMI-Informal Report 3/80*

Report for the TMI Special Inquiry Group. Analysis of the technical aspects of the factors associated with the accident at TMI.

NUREG-1362; *Data Summaries of Licensee Event Reports of Diesel Generators at US Commercial Nuclear Power Plants; 1/1/76-12/31/76 3/80*

Same procedure as described in 1205.

NUREG/CR-1363; *Data Summaries of Licensee Event Reports of Valves at US Commercial Nuclear Power Plants; 1/1/76-12/31/78 6/80*
Vol. I Report; Vol. II Appendices A-N; Vol. III Appendices O-Y

Same procedure as described in 1205.

NUREG/CR-1481; *Financing Strategies for Nuclear Power Plant Decommissioning 7/80*

"The report analyzes several alternatives for financing the decommissioning of N power plants from the point of view of assurance, cost, equity, and other criteria. Sensitivity analyses are performed on several important variables and possible compacts on representative companies' rates are discussed and illustrated."

NUREG/CR-1498; *Population Dose Commitments Due to Radioactive Releases From Nuclear Power Plant Sites in 1977 10/80*

See 1125 for description of method employed. Again, data for TMI is included.

NUREG/CR-1584; *Psychological Stress for Alternatives of Decontamination of TMI-2 Reactor Building Group* 8/80

"The purpose of this report is to consider the nature and level of psychological stress that may be associated with each of several alternatives for decontamination. The report briefly reviews some of the literature on stress, response to major disaster of life stressors, provides opinion on each decontamination alternative, and considers possible mitigative actions to reduce psychological stress.

"The report concludes that any procedure that is adapted for the decontamination of the reactor building atmosphere will result in some psychological stress. The stress, however, should abate as contamination is reduced and uncertainity is diminished. The advantages of the purge alternative are the rapid completion of the decontamination and the consequent elimination of future uncontrolled release. Severe stress effects are less likely if the duration of stress or exposure is reduced, if the feeling of public control is increased and if the degree of perceived safety is increased."

NUREG/CR-1620; *Survey of Current State Radiological Emergency Response Capabilities for Transportation Related Incidents* 9/80

"This volume is the final report of a project to survey current state radiological emergency response capabilities for transportation related incidents. The survey was performed to provide NRC with information useful in the development of guidelines for state organizations and planning for emergency response. The report includes the results of a mail and telephone survey of state emergency response officials; information gleaned from radiological emergency response plans and related official documents; and some general conclusions and recommendations drawn in part from interviews conducted and site visits to selected states."

NUREG/CR-1656; *Utility Management and Technical Resources* 9/80

ONRR contracted with Teknekron Research, Inc. to analyze and evaluate utility management and technical resources for dealing with events like that at TMI Unit II. Teknekron (1) analyzed licensee submittals in response to an NRC request to identify management and technical short-term and long-term resources for reacting to TMI II type accidents (2) developed acceptance criteria that specify minimum management and technical resources (onsite and offsite). The general conclusion is that those resources described in licensee submittals may be capable of dealing with TMI II type accidents.

NUREG/CR-1635; *Nuclear Plant Reliability Data System 1979 Annual Reports of Cumulative System and Component Reliability* 9/80

This report provides generic reliability information on systems and components for the cumulative period from 7/74 through 12/79. The reports include information about General Electric, B&W, Combustion Engineering, and Westinghouse units.

NUREG-1728; *The Feasibility of Epidemiologic Investigations of the Health Effects of Low-Level Ionizing Radiation* 11/80

This is a final report of "a study to determine the feasibility of conducting epidemiologic investigations of the health effects of low-level ionizing radiation," begun 7/3/79. The study defines low-level ionizing radiation as a single dose of 5 rem (whole-body) or less and chronic doses that accumulate at the rate of less than 5 rem per year.

"The objective of this project was to determine whether or not further epidemiologic research (either expansion of current projects or initiation of new ones) would be useful at this time for quantitating the health effects due to low-level ionizing radiation.

"No outstanding candidate population is recommended for study since, even if the largest available populations are studied, the chance of finding a definite positive result is very small.

However, the decision to conduct a study must rest heavily on social and political considerations rather than on purely scientific ones. Therefore, four populations are tentatively proposed for prospective cohort studies, with nested case-control studies as needed. Overall, the most practical approach would be to conduct a study through a national worker registry, with cancer as the endpoint of interest."

NUREG-1730; *Data Summaries of Licensee Event Reports of Primary Containment Penetrations at US Commercial Nuclear Power Plants; 1/1/72-12/31/78 9/80*

Same procedure as described in 1215.

NUREG-1779; *Emergency Preparedness for Nuclear Electric Generating Facilities in Foreign Countries: A Brief Survey of Practices 12/80*

"The report summarizes the emergency plans for accidents at N power plants in Germany, Sweden, Switzerland, the United Kingdom, Canada, and France. Plans were published before the accident at TMI occurred. Therefore, few ideas of immediate use to US planners.

"The study also discusses the emergency action levels, warning systems, evacuation management and procedures and public information and education for people living near power plants, and defines roles of N facility operators and roles of the government."

NUREG-75/087; *Standard Review Plan for the Review of Safety Analysis Reports for Nuclear Power Plants: LWR Edition 5/80*

The SRP is prepared for the guidance of staff reviewers in the ONRR in performing safety reviews of applicants to construct or operate N power plants. Each SRP describes the areas of review, acceptance criteria, review procedures, evaluation findings, and references.

B. Kemeny Commission Report

The President's Commission on the Accident at TMI, better known as the Kemeny Commission, was established by Pres. Carter in 4/79 to investigate the accident at TMI. The Commission issued its report in 10/79 and below are annotated the overall report as well as those of the various task forces.

Report of the President's Commission on the Accident at TMI Titled The Need for Change: The Legacy of TMI 10/79 Washington, D.C., John G. Kemeny, Chairman Reprinted by Pergamon Press, NY

1. The report includes the preface, overview, commission findings, commission recommendations, supplemental views and account of the accident. The appendices include the executive order, commission operations and methodology, commissioner's biographies, staff list, and glossary.
2. The overall conclusion was that "to prevent N accidents as serious as TMI, fundamental changes will be necessary in the organization, procedures and practices-and above all-in the attitudes of the NRC and to the extent that the institutions we investigated are typical of the N industry."

Technical Staff Analysis Report on WASH 1400 Reactor Safety Study

"The study determined that N accident risk is small, almost negligible, compared to more common risks, including airplane accidents, fires, dam failures, chlorine spills, earthquakes, hurricanes, and tornadoes. Results show that risk is small because the more likely reactor accident involves success of safety systems designed to accomodate them, and because accidents involving failure of safety systems are unlikely. These systems are provided in N reactors to prevent core meltdown and to diminish radioactivity release."

Technical Staff Analysis Report on Core Damage to President's Commission on the Accident at TMI

Selected findings were:

1. Ninety percent or more of the fuel rods have burst.
2. "In the hottest portions of the core, sections of the control rods probably melted. Because the constituents of the rods are essentially insoluble in water, the neutron poisons the rods contained are very likely still in the core."

Technical Staff Analysis Report on Chemistry to President's Commission on the Accident at TMI

Primary findings were:

1. "WASH-1400 (Rasmussen Report) concludes that an explosion, or detonation, within the containment building of the type approved by NRC of all hydrogen capable of being produced from the reactor's zirconium would not violate its ability to contain. An independent assessment by LASL found that the containment building at TMI II may be marginal in its ability to withstand such a deterioration of all the hydrogen produced.
2. "The hydrogen production rate at TMI II was of the order of 500 times the capacity of the existing recombiner."

Technical Staff Analysis Report on Condensate Polisher to President's Commission on the Accident at TMI

The findings were as follows:

1. The condensate polisher effluent values closed at the beginning of the accident on 3/28/79.
2. Tests at TMI II, to date, have not confirmed a reason for closure.

The primary conclusion was that the "condensate polisher, though vital to the operation of the plant, did not receive appropriate attention in design and attention from assurance function, engineering management review groups; proper attention could provide a significant increase in plant reliability."

Report of the Office of Chief Counsel on the NRC

This report presents the historical context of Energy Reorganization Act of 1974; Post 1974 Structure of the NRC and the role of the Commissioners; Plant Licensing; Inspection and Enforcement; Operator Licensing and Training; and finally, NRC Emergency Response.

The report concluded that the communication system was very poor in that there was no direct link between the site and the control area.

Report of the Task Group on Health Physics and Dosimetry to President's Commission on the Accident at TMI

The primary task of the group was to determine the radiation doses that the worker population and the general public within a 50 mile radius of TMI received as a result of the accident. Estimates are presented for both groups including the premises used to reach these estimates. Finally, the task group found that the "maintenance of instruments and housekeeping (at TMI) were below the standards for a good health physics program."

Reports of the Office of Chief Counsel on Emergency Preparedness, Emergency Response

The report examines the status of emergency planning at the time of the TMI accident. It reviews NRC's requirements on reactor siting and emergency planning. It examines federal, state, and local planning activities for emergency preparedness prior to the accident.

The major conclusion is that the "NRC regulatory approach and the lack of urgency with which various levels of government have conducted planning activities indicates a fundamental problem of attitude that is woven into the fabric of the radiological emergency planning in place at the time of the TMI accident."

Report of the Emergency Preparedness and Response Task Force

The report examines the planning that existed at the time of the TMI accident in the counties surrounding the plant and at the state and federal levels. It also examines the responses of the various governmental units following the start of the accident.

This report includes a discussion of the methodology utilized and three appendices are available. They are:

a. Analysis of Emergency Plans
b. The role of NRC in Emergency Plans and Response
c. Analysis of Evaucation Behavior

Reports of the Public Health and Safety Task Force

This report summarizes the findings of the Health Physics and Dosimetry Task Group, the Radiation Health Effects Task Group, Behavioral Effects Task Group, and the Public Health and Epidemiology Task Group. It also presents each report in greater detail.

Technical Assessment Task Force on Control Room Design and Performance

"There is evidence that the operators of TMI II were confused by equipment indications available to them on 3/28/79. During the course of the accident which began that day, a number of malfunctions of control equipment occurred. This complicated the problem operators were facing or caused additional confusion. For this reason, the control room design was reviewed to evaluate both its adequacy in providing the necessary information to operators and the controls needed to shutdown the plant and maintain it in a safe condition. Performance of the control room during the transient was assessed. Finally, industry efforts to improve control room design through 'human factors engineering' were reviewed."

Technical Staff Analysis Report on Transport of Radioactivity From the TMI-2 Core to the Environs

The primary findings are that 1) the discharges of pressure relief systems that communicate with the primary coolant were not routed to the reactor containment system. 2) The reactor containment building is not isolated on radiation signals. This isolation probably would not have precluded the airborne radioactive releases at TMI II. 3) It appears that inadequate attention was given to design to assure that the water traps WDG-U8A and WDG-U9A had lower pressure capabilities than the pressure relief valve WDG-R-3 of the vent header system. 4) The concrete in the auxiliary building was not sealed prior to startup. 5) Readily accessible up-to-date readable drawings and specifications were not available on site.

Technical Staff Analysis Report on Quality Assurance

The major findings are as follows:

1. "The NRC organization, procedures and practices, as now constituted, do not provide for the combined management, engineering, and assurance review of utility performance necessary to minimize the probability of equipment and operator failures and necessary to ensure the safe operation of the N plant.
2. A lack of independent onsite quality assurance of safety assessment of plant operations and of equipment not considered 'safety related' contributed significantly to the accident at TMI.
3. There was a lack of detailed safety and failure modes analysis on all plant systems necessary to ensure the reliability and safety of the facility.
4. Current utility and NRC practices do not assure proper preparation, review, and execution of operating and maintenance procedures.
5. NRC has very limited view of changes made to plant configuration. Utility control of safety related equipment changes appears adequate; control of non-safety related equipment configuration is inadequate.
6. Full use is not being made of management, engineering, safety, reliability, and quality assurance practices which are in use in other industries where safety and reliability are critical concerns."

Technical Staff Analysis Report on Iodine Filter Performance

Findings:

1. "During the accident at TMI a quantity of Iodine 131 was detected in the gaseous effluent. This quantity was more than that which would be expected to pass through the filtering system if it performed as designed. Replacement charcoal in the auxiliary building ventilation system and in half of the fuel handling

building ventilation system significantly reduced the iodine discharges suggesting that charcoal in the filter trains at the onset of the accident did not perform as expected.

2. The air filtering systems were designed to be used only when needed to remove airborne radioactivity, because of a limited filtering lifetime for charcoal. However, ventilation flow had been through the filters for about one year. This fact coupled with the lack of surveillance to verify system performance could explain apparently inadequate filter performance during the accident."

Technical Staff Analysis Report on Closed Emergency Feedwater Values

Summary

On 3/28/79, the TMI II N power plant experienced the most severe accident in the U.S. commercial N power plant operating history. The accident which occurred at about 4:00 in the morning was started by a loss of normal feedwater supply to the steam generators, which led rapidly through a normal sequence of events for reactor shutdown. During this normal sequence of events, The PORV opened and stuck (failed) in the open position in which it remained for a considerable length of time before being isolated by action of the operators. This failure followed by early operator action that throttled the flow from the high pressure injection pumps, initiated an abnormal sequence of events that led to this most severe accident.

A number of analyses, involving studies, inspections, and tests, have been conducted to understand what caused this accident and why it happened. One of these analyses was to investigate the reason for the emergency feedwater valves being in the closed position instead of the open position as required, as described in this report.

The findings and conclusions from this analysis are as follows:

1. There has been no positive identification of a reason for the valves being in the closed position.

2. Of all the explanations analyzed, the most likely explanations are:
    a. the valves were not reopened at the conclusion of the most recent surveillance procedure, requiring them to be closed, conducted prior to the accident;
    b. the valves may have been mistakenly closed by control room operators during the very first part of the accident;
    c. the valves may have been mistakenly closed from other control points within the plant; and
    d. while considered a remote possibility, there is a chance that these valves were closed by an overt act.
3. The utility failed to apply appropriate control over safety related procedure and its implementation and changed to it; NRC failed to detect lack of control.
4. The utility does not apply appropriate discipline to access to in-plant areas, accomplishment of procedures, and equipment configuration. NRC did not recognize this lack of discipline.

Technical Staff Analysis Report on Behavioral Effects

The report presents the findings based on surveys of about 2,500 persons from four different groups. They are 1) the general population of male and female heads of households located within 20 miles of TMI; 2) mothers of preschool children from the same area and a similarly drawn control sample from Wilkes-Barre which is about 90 miles away; 3) teenagers in the 7th, 9th, and 11th grades from a school district within the 20 mile radius of TMI; and 4) workers employed at TMI at the time of the accident and a control group of workers from the Peach Bottom nuclear plant about 40 miles away.

"In brief, the TMI accident, had a pronounced demoralizing effect on the general population of the TMI area, including its teenagers and mothers of preschool children. However, this effect proved transient in all groups studied except the workers, who continue to show relatively high levels of demoralization. Moreover, the groups in the general population and the workers, in their different ways, have continuing problems of trust that stem directly from the accident. For both the workers and general population, the mental health and behavioral effects are comprehensible in terms of the objective realities of the threats they faced."

Task Force on Pilot Operated Relief Valve

The primary findings are:

1. The PORV apparently failed in the open position at TMI II on 3/28/79; TMI operators had no positive indication of the open/close position of the PORV; the absence of this signal in the control room contributed to the confusion of the operator during the TMI II accident.
2. Failure of the PORV in the open position results in a small-break LOCA.
3. Existing procedures did not consider a stuck-open PORV as a small-break LOCA.
4. The PORV was not classed as a safety related component of the reactor coolant system.

The major conclusion was that the TMI II accident would probably not have progressed beyond a severe feedwater transient, had the PORV been recognized and treated as a safety related component.

Technical Staff Analysis Report on Selection, Training, Qualification of TMI Reactor Operating Personnel

The fundamental problems related to the training and licensing of TMI operators might be highlighted as follows:

1. There was a gulf between the operators or operationally-oriented personnel and the managers or other decision makers in the NRC, at Met Ed, and at B&W. Few communications took place between B&W management/engineering and simulator instructors. The NRC Operator Licensing Branch, which set and enforced the standards for operator training, was understaffed and lacked outside direction. The TMI management did not consider itself responsible for operator training.
2. The training standard was low and did not require that the operators be provided with

the analytical tools necessary to operate a nuclear reactor.

3. There was no effective mechanism for learning from the mistakes of others. The system was such that, in large part, the accident at TMI would not have occurred if the operators had been thoroughly and comprehensively trained on the lessons of Davis-Besse.
4. There was no effective mechanism for ensuring a high level of knowledge. No competent outside organization periodically determined in-depth operator knowledge nor did the licensing and requalification process accomplish this. Unless the utility was enlightened and had the resources to ensure its own high standard then its operators might attain only mediocre knowledge and skill.
5. There was no consideration given to training engineers at a higher level than the reactor operators. This stemmed, perhaps, from the underlying assumption that a nuclear reactor which produces power from a highly complex process and has the potential for affecting the health and welfare of the public can be operated solely by a few high school graduated or "equivalent."

Report of the Office of Chief Counsel on the Role of the Managing Utility and Its Suppliers

The primary findings are as follows:

1. Operating experience at TMI II and other N power plants was not being effectively gathered, analyzed, or followed up.
2. Training of operators and management personnel was seriously inadequate and did not stress fundamentals of reactor safety.
3. Although plant procedures went through a multiple-step review process, there were significant deficiences in the procedures in use at the time of the accident.
4. There was insufficient transfer of the knowledge gained by GPU in the design and construction of the plant to the Met Ed personnel responsible for operating the plant.

Technical Staff Analysis Report on the Public's Right to Information

The primary findings are:

1. "The public information problems of Met Ed and the NRC were rooted in a lack of planning. Neither expected that an accident of this magnitude-one that went on for days, requiring evacuation planning-would ever happen. In a sense they were victims of their own reassurances about the safety of nuclear power. As a result, neither had a 'disaster' public relations plan. Coordination between the utility and the NRC was so weak that responsibility for informing the public in the first crucial hours of the accident was undefined. The NRC did not know when, or whether, to send its own public information people to the site, or when, or where to set up a NRC press center.
2. Perhaps the most serious failure in the planning stage was that neither the utility nor the NRC made provision for getting information from the people who had it (in the control room and at the site) to the people who needed it.
3. The news media were also somewhat unprepared, and this added to the prevailing confusion. While it is a goal of many journalistic organizations to develop specialists who are expert in particular areas (such as business reporting, science and medical reporting, or national political reporting), few reporters who covered TMI had more than rudimentary knowledge of nuclear power. Some, by their own admission, did not know how a pressurized, light water reactor worked, or what meltdown was. Few knew what questions to ask about radiation releases so that their reports could help the public evaluate health risks."

C. Federal Government Documents

There are two major sources for government documents. They are the monthly catalog of U.S. Government Publications and the CIS Index. In both cases, the publications contain both a table index and a specific item abstract. Furthermore, in both cases, there is no title index such as TMI until 1979. For TMI items prior to 1979, the reader has to utilize other title indexes, items such as atomic power, the AEC, and the NRC.

The Monthly Catalog of U.S. Government Publications is issued by the Superintendent of Documents, U.S. Government Printing Office, Washington, D.C. In each catalog, there is an explanation of the numbering system employed by them. This publication is usually available in those libraries serving as government documents depositories. They also may be available at other libraries.

1979 TMI Items

Accident at the Three Mile Island Nuclear Power Plant: Oversight Hearings Before a Task Force of the Subcommittee on Energy and the Environment of the Committee on Interior and Insular Affairs, House of Representatives, Ninety-sixth Congress, First Session..., 79-17508.

Evaluation of Long-Term Post-Accident Core Cooling of Three Mile Island Unit 2: NRC staff report/, 79-15819.

Investigation Into the March 28, 1979, Three Mile Island Accident/, 79-17354

Population Dose and Health Impact of the Accident at the Three Mile Island Nuclear Station: (A Preliminary Assessment for the Period March 28 Through April 7, 1979)/, 79-21289.

Population Dose and Health Impact of the Accident at the Three Mile Island Nuclear Station: Preliminary Estimates for the Period March 28, 1979 Through April 7, 1979/, 79-21301

Summary and Discussion of Findings From Population Dose and Health Impact of the Accident at the Three Mile Island Nuclear Station: (A Preliminary Estimate for the Period March 28 Through April 7, 1979)/, 79-20401.

TMI-2 Lessons Learned Task Force Status Report and Short-Term Recommendations, 79-21308

Title List Publicly Available Documents Three Mile Island Unit 2 Docket 50-320: Cumulated to June 30, 1979/, 79-21305.

Title List Publicly Available Documents Three Mile Island Unit 2 Docket 50-320: Cumulated to May 21, 1979/, 79-21304.

1980 TMI Items

Accident at the Three Mile Island Nuclear Power Plant: Oversight Hearings Before a Task Force of the Subcommittee on Energy and the Environment of the Committee on Interior and Insular Affairs, House of Representatives, Ninety-sixth Congress, First Session..., 80-9306, 80-11136.

Civil Defense and the Three Mile Island Nuclear Accident: Report of the Military Installations and Facilities Subcommittee of the Committee on Armed Services House of Representatives, Ninety-sixth Congress, First Session, December 18, 1979, 80-19346.

Civil Defense Aspects of the Three Mile Island Nuclear Accident: Hearings Before the Military Installations and Facilities Subcommittee of the Committee on Armed Services, House of Representatives, Ninety-sixth Congress, First Session, 80-9186.

Computer-Plotted Map of Land Use and Land Cover, Three Mile Island and Vicinity, With Census Tracts, 80-18860

Generation of Hydrogen During the First Three Hours of the Three Mile Island Accident/, 80-5408.

Human Factors Evaluation of Control Room Design and Operator Performance at Three Mile Island-2: Final Report/, 80-9144.

Impact Abroad of the Accident at the Three Mile Island Nuclear Power Plant: March-September 1979/, 80-26693.

Kemeny Commission Findings: Oversight Hearing Before the Subcommittee on Energy Research and Production of the Committee on Science and Technology, U.S. House of Representatives, Ninety-sixth Congress, First Session, November 14, 1979, 80-15741.

Known Effects of Low-Level Radiation Exposure: Health Implications of the TMI Accident, April 1979/, 80-14679.

NRC's Response to the Report of the President's Commission on Three Mile Island: Hearing Before the Subcommittee on Energy and Power of the Committee on Interstate and Foreign Commerce, House of Representatives, Ninety-sixth Congress, First Session, November 5, 1979, 80-19466.

Nuclear Power Plant Safety After Three Mile Island: Report/, 80-19529.

Radiation Measurements Following the Three Mile Island Reactor Accident/, 80-16395.

Report of the Emergency Preparedness and Response Task Force/, 80-12631.

Report of the Office of Chief Counsel on Emergency Preparedness Emergency Response/, 80-8725.

Report of the Office of Chief Counsel on the Nuclear Regulatory Commission/, 80-8726.

Report of the Office of Chief Counsel on the Role of the Managing Utility and Its Suppliers/, 80-8727.

Report of the President's Commission on the Accident at Three Mile Island: The Need For Change: The Legacy of TMI, 80-6415.

Report of the President's Commission on the Three Mile Island Accident: Joint Hearing Before the Subcommittee on Nuclear Regulation of the Committee on Environment and Public Works, United States Senate, and Subcommittee on Energy and the Environment of the Committee on Interior and Insular Affairs, House of Representatives,

Ninety-sixth Congress, First Session, October 31, 1979, 80-15728.

Report of the Public's Right to Information Task Force/, 80-12633.

Reports of the Public Health and Safety Task Force on Public Health and Safety Summary, Health Physics and Dosimetry, Radiation Health Effects, Behavioral Effects, Public Health and Epidemiology/, 80-12632.

Supplemental Expenditures by the Committee on Environment and Public Works: Report to Accompany S. Res. 171, 80-12903.

Three Mile Island: A Report to the Commissioners and to the Public/, 80-13100, 80-15452.

Three Mile Island Nuclear Accident, 1979: Hearing Before the Subcommittee on Health and Scientific Research of the Committee on Labor and Human Resources, Unites States Senate, Ninety-sixth Congress, First Session ... April 4, 1979, 80-9318.

Three Mile Island Nuclear Plant Accident: Hearing Before the Subcommittee on Natural Resources and Environment of the Committee on Science and Technology, U.S. House of Representatives, Ninety-sixth Congress, First Session ... June 2, 1979, 80-13324.

Three Mile Island Nuclear Power Plant Accident: Hearings Before the Subcommittee on Nuclear Regulation of the Committee on Environment and Public Works, United States Senate, Ninety-sixth Congress, First Session..., 80-9336, 80-11178, 80-13304.

Three Mile Island Telephone Survey: Preliminary Report on Procedures and Findings/, 80-3520.

TMI-2 Lessons Learned Task Force Final Report, 80-3465.

1981 TMI Items

Answers to Questions About Removing Krypton From The Three Mile Island Unit 2 Reactor Building/, 81-2961.

Environmental Assessment of Radiological Effluents From Data Gathering and Maintenance Operation on Three Mile Island Unit 2: Interim Criteria Approved by the Commission on April 7, 1980/, 81-2962.

Most Studied Nuclear Accident in History Summary: Report to Congress, 81-1483.

NRC Plan for Cleanup Operations at Three Mile Island Unit 2/, 81-2964.

Three Mile Island Cleanup and Rehabilitation: Oversight Hearing Before the Subcommittee on Energy and the Environment of the Committee on Interior and Insular Affairs, House of Representatives, Ninety-sixth Congress, Second Session, on Three Mile Island Cleanup and Rehabilitation Hearing Held in Washington, D.C. May 22, 1980, 81-3078.

Related Requirements for New Operating Licenses: 81-1032.

The CIS Index lists congressional publications including testimony and public laws. The first volume appeared in 1970 and their address is 4520 East-West Highway, Washington, D.C. 20014. The CIS publications are usually available in those libraries serving as government document depositories. They also may be available at other libraries.

In relation to TMI, the index items appear with their access in numbering system; an explanation of this system appears in the introduction to each volume.

1979

Accident, H 441-18, H 441-26, S 321-24
Accident, health implications, H 701-53.1
Accident, impact on US nuclear energy future, H 181-96.1

Decontamination Advisory Panel, estab., H 503-33
DOE Nuclear Technology R&D Programs fy 81 approp., H 181-36.3, S 181-21.3
Electric Power Research Inst. Study, H 701-9.4
Health Effects of Low-Level Radiation Exposure, Fed. Research and Safety Programs, S 401-5.2
NRC, budget review fy 81, S 321-24.3
NRC, mgmt. improvements, pres. message, H 400-1
NRC, organizational improvements implemented since Three Mile Island incident, S 401-90.4
NRC, programs fy 81 and supplemental fy 80 approp., S 181-18.4
NRC, programs fy 81 approp., H 181-35.1, H 181-73.3
NRC, programs fy 81 authorization, H 441-37, H 501-77
NRC, programs, fy 81 authorization and response to Three Mile Island nuclear power plant accident, S 321-25
NRC, programs, supplemental approp. fy 80, H 181-53.14
NRC, Spec. Inquiry Group on Three Mile Island Recommendation Review, H 401-57
Nuclear Power Plant Accident Liability Coverage, H 441-16
Nuclear Power Plant Accidents, Emergency Preparedness Planning, H 401-24
Nuclear Power Plant Radiation Monitoring, Emergency Planning, and Siting Issues, S 401-6
Radiation Monitoring Activities by Fed. Agencies, H 701-9.1
Safety Management Factors Germane to the Nuclear Reactor Accident at Three Mile Island, S 401-90
Significance of Three Mile Island, Commonwealth Edison Co. Analysis, H 261-5.3

1981 (January-March)

Nuclear Power Plant Accidents, Emergency Preparedness Planning, 2 H 441-2.5

D. Pennsylvania Publications

These state government publications are available at the State Library of Pa., Commonwealth and Walnut Streets, Harrisburg, PA 17105. In relation to TMI, they are organized in the following areas: Commission to Study and Evaluate the Consequences of the Incident at TMI, Department of Health, Department of Military Affairs, and the Governor's Office. For each document, the State Library Call Numbers are presented.

Commission

1. *A Layman's Guide to Radiological Health Information and Resources: A Selected Bibliography* (Harrisburg) 1980, PYT 5312.2, L4279.

2. *Minutes of Meetings of Various Subcommittees as well as Commission as a Whole* (Harrisburg), PYT 5312.14/6.

3. (News Releases), PYT 5312.15/4.

4. *Report: May 14, 1979 and Executive Order No. 1979-3*, PYT 5312.2, C734.

5. *Report in Response to NRC Staff Recommended Requirements for Restart of TMI Unit #1-Met Ed/GPU*, PYT 5312.2, R311.

6. *Report of the Governor's Commission on TMI*, Harrisburg, 1980, PYT 5312.2, R3119.

7. *Report of the Legal Subcommittee of the Governor's Commission on TMI*, Harrisburg, 1979, PYT 5312.2, L496a.

Department of Health

1. *Health-Related Behavioral Impact of the Three Mile Island Nuclear Incident:* Report Submitted to the TMI Advisory Panel on Health Research Studies of the Pennsylvania Department of Health by Peter S. Houts, Principal Investigator ... (et al.), (Hershey), Pennsylvania State University, College of Medicine, 1980, PHE 1.2, H434b.

2. *Health-Related Economic Costs of the Three Mile Island Accident:* A Progress Report Submitted to the Division of Epidemiological Research, Teh-wei Hu, Principal Investigator ... (et al.), University Park, PA, Pennsylvania State University, 1980, PHE 68.2, H434e.

Department of Military Affairs

1. *Three Mile Island Nuclear Incident 28 March-5 April 1979*, Adjutant-General's Office, The Pennsylvania National Guard, Annville, PA, 1979, PMA 23.2, T531.

2. *Nuclear Reactor Accidents-1979; After Action Report*, OCLC 5229062.

Governor's Office

Office of Policy and Planning

The Socio-Economic Impacts of the Three Mile Island Accident: Final Report, Prepared in cooperation with: the Department of Agriculture, the Department of Community Affairs, the Department of Commerce, the Department of Insurance, the Department of Labor & Industry the Department of Revenue, Harrisburg (1980), PYP 712.2, T531mf.

Office of Policy and Planning

Three Mile Island Socio-Economic Impact Study, Prepared in cooperation with: the Department of Agriculture, the Department of Community Affairs, the Department of Commerce, the Department of Insurance, the Department of Labor & Industry, the Department of Revenue, Harrisburg (1979), PYP 712.2, T531m.

PROFESSIONAL PAPERS, ARTICLES, AND BOOKS DEALING WITH THREE MILE ISLAND

In this section, the annotations of professional papers and articles as well as the listing of books dealing with TMI are presented.

Despite efforts to solicit current publications on the topic, there was limited response to our request. Therefore, the number of papers annotated represent only those submitted to the authors or secured by them.

Once again, there was a lack of information prior to the crisis at TMI in 1979. Professional interests tend to mirror popular concerns.

1. Cutter, Susan L. and Barnes, Kent, "Three Mile Island: Risk Assessment and Coping Responses of Local Residents." Rutgers University Geography Discussion Papers, January, 1982.

   This study reports on questionnaire surveys of TMI area residents both one month and one year after the accident. Basically the researchers are interested in reactions to the accident and steps taken by residents to protect themselves from further risks.

   Topics reported from the initial survey include: speed of information transmission, perception of danger during crisis, reliability of information, assessment of future risk, trust in institutions, reasons for evacuation, and factors influencing the evacuation decision.

   In the year later, follow-up respondents were again queried as to their assessment of risk of future accidents. Their assessment continued to be low. Interestingly, their assessments were manipulated by presenting both a general risk and a specific, i.e., TMI risk. Data is also presented on how people would act differently if the situation would reoccur.

2. "Mental Health Effects of the Nuclear Accident at Three Mile Island." A symposium presented at the Annual Meeting of the American Psychological Association, September, 1980, Montreal, Canada.

The purpose of the symposium is "to assess behavioral and mental health effects of the accident" and to understand the implications of varied research studies. The goal is to analyze the need for energy versus the need for health and safety.

The first of the four abstracts by G.S. Warheit, is entitled "Contextual Environment of the TMI Nuclear Accident: A Socioanthropological Overview." The area, the people, the social psychological environment, and responses to the disaster are investigated.

The second abstract by Barbara S. Dohrenwend, Bruce P. Dohrenwend, and Raymond L. Goldstein, studies the "Impact on Household Heads and Mothers of Preschool Children." The results of the study show that household heads living within five miles of the plant and mothers of preschool children were "the most demoralized persons."

The abstract of "Reaction of Adolescents to the Emergency at Three Mile Island"-a survey done by F.S. Barlett, J.L. Martin, and L. Byrnes, describes a study made of students in grades 7, 9, and 11 in the school district in which TMI is situated. Questionnaires completed 2 months after the accident by 632 students were evaluated, and the results indicate that, "in general, the students were quite upset ... by the circumstances." The degree of distress experienced by the adolescents was determined to be related to the presence of preschool age siblings, residential proximity to TMI, and other risk and stress factors.

The final abstract contained in the paper reviews the survey prepared by S.V. Kasl, and R. Chisholm, entitled "Impacts of the TMI Accident on Nuclear Workers." The study was performed on fulltime TMI employees and workers at the Peach Bottom plant. The data were obtained through telephone interviews. Findings include behavioral and mental health effects, and this information is compared to data collected from other groups included in the overall study supported by the Kemeny Commission.

The conclusions advise that the workers and general population affected by TMI exhibit "continuing problems of trust that stemmed directly from the accident." Recommendations concerning public policy and additional research are included.

3. "Comments at Symposium on Interface of Hardware and Human Factors-Three Mile Island," J.B. Brilliant, presented at American Psychological Association Annual Meeting, September, 1980, Montreal, Canada.

This paper studies the problems that have surfaced with the extensive use of computers for communication and information processing. The author specifically deals with the problem that organizations face of purchasing unnecessary and inappropriate equipment to serve their needs. Neglecting to assess both the organization's information needs and the actual contribution made by the information system to the organization's productivity are cited as main reasons for the problem.

Brilliant points out that the current trend is to emphasize "quantity of information and the quality of the physical storage and distribution subsystems." It is alleged that information technologists disregard the human factors while selecting a system, placing chief importance on the sophistication of the hardware and the amount of information.

The paper discusses a feature of information-uncertainty. The hazards of N radiation are used as examples that illustrate the impact of uncertainty in information. The fact that there is no universally accepted determination of danger for some threshold of radiation is a source of uncertainty.

The paper briefly reviews the conditions during the accident at TMI and advises that communication during events such as those that occurred at TMI can be improved by improving the relationship between the human, the information, and the physical communications facilities.

4. "The Credibility of Government and Utility Officials in the Aftermath of Three Mile Island," Raymond Goldstein, John K. Schorr, and John Martin. Paper presented to PSS, November, 1979.

A report on the results of questions asked of four populations in the TMI area. The categories include the general population of the TMI area, the general population of the TMI area subdivided by sex, the population of mothers of young children in the TMI area, and the population of mothers of young children in Wilkes-Barre.

Several questions concerning the credibility of the officials involved in the TMI events were asked of these groups. The results are evaluated and then interpreted on a "trust scale."

5. "Demographic and Attitudinal Characteristics of T.M.I. Evacuees," Donald Kraybill, Daniel Buckley, and Rick Zmuda. Paper presented to PSS, November, 1979.

This paper reports the differences in responses between evacuees and nonevacuees. The evacuation behavior is reported in five tables. These tables classify the respondents' evacuation behavior according to sex, age, education, household size, and distance from TMI.

After the demographic study, the article goes on to report the responses given in answer to attitude-oriented questions according to the five aforementioned categories. The paper concludes with separate summaries of the demographic and attitudinal conclusions suggested by the offered evidence.

6. "Why The TMI Disaster Is Different," Philip Starr, Paper Presented to PSS, November, 1979.

This paper explains why the TMI mishap differs from other disasters. Starr begins by giving a brief synopsis of the events at TMI. He states that TMI is unique among disasters "because there was no agreement as to: 1) whether a disaster occurred, 2) whether it could be controlled, 3) whether the area could be safely evacuated, and 4) whether the events at TMI are still a health and safety threat to local residents."

Starr also includes a short description of the four social implications he views as valid reasons for research.

7. "Innocence Lost," Mike Vargo, *Pennsylvania Illustrated*, Vol. 3, #6, August, 1979, pp. 23-64.

This lengthy article zeroes in on a select few of Middletown's residents. Bob Reid (the mayor) and John Garver (a man described as a "river rat") and his family are dealt with in great detail.

The article concerns itself primarily with the reactions and personal feelings of Middletown residents during the TMI mishap. Each new development at the plant and the subsequent actions taken by Middletowners are described on a very personal level. The descriptions of the many reporters, the NRC members, and the many N experts at the scene are presented as seen through the eyes of the small town's inhabitants.

The confusion and fear of the unknown are vividly evident in the section devoted to Mayor Reid. The many factors that influenced his decisions concerning Middletown citizens are given in detail. His decisions to create an evacuation plan and to tighten security in order to prevent looting of vacated homes are mentioned as are similar decisions.

In general, the article is written in a story-like fashion and views the TMI incident from the Middletown point of view. It reports the TMI events from a very human level.

8. "Three Mile! The Accident Isn't Over," Lynne Shivers, *Christianity and Crisis*, November 12, 1979, pp. 274-286.

The "personal emotional impact" that the TMI event had on local residents is discussed. A Middletown physician (John Barnoski) lists "fatigue, malaise, nervousness, insomnia" and other related disorders as symptoms resultant of the TMI accident.

The article goes on to explain some of the findings of Dr. Robert J. Lifton, a physician whose studies deal primarily with "psychological stress resulting from traumatic events." Five emotional responses observed by Lifton in people who have suffered trauma are defined and applied to citizens who have been affected by the TMI accident.

The table entitled "Evacuation Behavior by Interview Variables" presents the responses in percentages elicited from a population interviewed on TMI. The questions asked were primarily concerned with the information gathering methods of the population and the evacuation decision.

9. "Nuclear Energy," A *Lancaster Sunday News* Public Opinion Poll, conducted by Donald B. Kraybill and Raymond K. Powell, December 11, 1979.

This is a public opinion poll conducted by the SRC at the request of the *Lancaster Sunday News*. The poll was conducted during the period of 11/26 to 12/4/79.

After a brief introduction, the report carefully explains the methods used in obtaining a random sample of Lancaster telephone numbers. The response rate, age, and sex distributions and findings are also given. Included in the findings section is an explanation of the sampling error inherent in the poll.

The body of the report consists of ten questions suggested by the *Sunday News* that were asked of the public regarding nuclear power and the TMI accident. The responses are recorded in percentages, and they are classified according to the age and sex of the respondent.

10. "The Legacy of Three Mile Island," *Syracuse University Alumni News*, Winter, 1981, Vol. 61, #2, pp. 13-19.

This article is primarily a review of the activities and studies of Syracuse University alumni and professors in the area of N energy. Specifically, the article reports the comments, opinions, and explanations of these scholars regarding the recent TMI accident.

A brief introduction of the three men described as "pioneers in N energy" -- Melvin Silliman, Roland Smith, and Raymond Pasternak is offered.

The accident on 3/28/79 at TMI is explained with reference to a book entitled *Three Mile Island: Prologue or Epilogue?* by Daniel W. Martin. Specific examples of misunderstandings, serious judgment errors, and scientific miscalculations are cited. The report goes on to outline a few major groups ordered to investigate the accident. Important names in the investigations are mentioned.

The article concludes with a list of new regulations effected since the investigations, a brief account of the N energy controversy, and a simple explanation of how N plants produce power.

11. "Resource Mobilization Theory and the Dynamics of Local Anti-Nuclear Coalition Formation in the Wake of the Three Mile Island Accident," Edward A. Walsh, paper presented at American Sociological Association Meetings, New York, August, 1980.

This lengthy paper discusses the activities of organized protest groups formed after the accident at TMI and reports the relationships between various anti-TMI organizations. Restricting its analysis to the first year following TMI, the report focuses on the resource mobilization theorists' neglect of the role of grievances in precipitating and sustaining the mobilization process.

The paper begins by defining terms such as "social movement organization," "social movement," "social movement industry," and "social movement coalition." It then goes on to discuss the Smelserian "generalized belief" model.

Four hypotheses central to the main discussion of the paper are presented. Two of the hypotheses consider the early precipitants of the mobilization process, and the second two deal with social movement organization interaction. Graphs, diagrams, comments from TMI area protestors, and other relevant data are presented as evidence to support the report's thesis.

12. "Three Mile Island Health Effects Research Program," George K. Tokuhata, *Proceedings of the Pennsylvania Academy of Science*, Vol. 54, Issue No. 1, 1980, 19-21.

The Director of the BER for the PDH describes 11 separate studies undertaken by the department either alone or in cooperation with other research organizations. These studies are also designed to measure and assess health and mental health outcomes associated with the TMI accident.

13. "Extent and Duration of Psychological Distress of Persons in the Vicinity of Three Mile Island," Peter S. Houts, Robert W. Miller, Kum Shik Ham, and George K. Tokuhata, *Proceedings of the Pennsylvania Academy of Science*, Vol. 54, Issue No. 1, 1980, 22-28.

This paper describes the crisis at TMI as "ongoing." It examines psychological responses of TMI area residents at the time of the crisis and at three and nine months later. Three aspects of the responses are described: 1) feelings of overall concern, 2) effects on self-reports of stress-related somatic and behavioral symptoms, and 3) effects on responses to the Lagner Index of Psychological Distress. Geographic proximity to TMI is a primary variable of analysis and comparison. Higher levels of stress are reported closer to TMI, attitudes also differ on the geographic proximity variable. Fifteen miles out from TMI seems to be a chance point in terms of attitudes, stress, and physical symptoms.

14. "Industry Response to Three Mile Island," Bruce L. King, *Proceedings of the Pennsylvania Academy of Science*, Vol. 54, Issue No. 1, 1980, 29-30.

This article describes changes in the organization of the N industry, higher standards of operator training, and new hardware devices that have been introduced as a result of the TMI accident.

15. "Information Types, Cognitive Heuristics, and Political Persuasion," by J.D. Winkler, S.E Taylor, R.F. Tebbets, J.B. Jemmott III, and J. Johnson, paper presented to the Annual Meeting of American Psychological Association, September, 1979, New York.

In order to examine the possibilities, the Bakke case and its newspaper coverage were monitored. Articles were classified into three groups -- case history information, base rate information, and anchoring information. It is argued that case history information may prove the more persuasive of the three. Case history information is then more carefully studied to determine whether vividly presented case history information is more persuasive than less vivid information.

The paper goes on to review two studies performed in order to arrive at a conclusion on this issue. The results of both studies showed that there was virtually no evidence that vivid communications were more impactful or persuasive than nonvivid ones.

One study, however, reflected an effect of vividness. Prior to the TMI events, subjects were asked to read one of two communications (one was vivid, the other nonvivid) and answer questions as to the expected frequency of nuclear accidents. After TMI, the subjects were asked to respond again to the materials. Pre-TMI data indicated that the vividness of the communication made little difference. Post-TMI data show other conclusions. Vivid communications produced more anxious responses in the subjects than did nonvivid materials.

The conclusion drawn from this study is that vivid information may not have a persuasive impact by itself; however, it "may do so in interaction with other variables such as an actual event ..."

16. "Harold Denton," Lorna Simmons Nolt, *Susquehanna Magazine*, Vol. 5, No. 2, February, 1980, pp. 12-13.

The article discusses specifically Denton's role as NRC spokesman at the TMI plant during the accident. The trust and respect he won from frightened area residents are attributed to his "straight forward, sure manner."

The author records Denton's responses to questions such as, "Why did the accident at TMI happen?" and "What is being done to remedy the apparent lack of crisis preparedness?" It concludes with a short synopsis of Denton's observations on the future of N energy, his plans and feelings for his job, and his family life.

17. "The Meltdown That Didn't Happen," Howard Morland, *Harper's*, Vol. 259, No. 1553, October, 1979, pp. 17-20.

This detailed account compares what did happen at TMI during the accident and what could have happened. The author states that "providence" and "not sound engineering and not good judgment" is to be thanked for the outcome of the incident.

The author begins be explaining the fission process in a single uranium atom and the by-product of two radioactive atoms. These atoms try to achieve N stability by rearranging themselves; this activity is termed "radio-active decay." Radioactive decay generates heat and cannot be stopped as can the fission reaction. Only the passage of time can turn off decay heat.

The TMI Unit II reactor contains a core consisting of a "delicate assembly of uranium fuel rods." It is stressed that the fuel rods must constantly be bathed with cooling water. An improperly cooled core can cause the rods to melt, and the devastating consequences of such an occurrence are outlined.

The author points out that "reactors cannot explode like atom bombs." But, he is quick to add, their fuel can "leave the buildings in which they reside and march across the country-side like the angel of death."

The bulk of the rest of the article concentrates on "hypothetical meltdowns" and the resulting highly dangerous radioactive plume. The TMI events are discussed with emphasis on the weather and hydrogen explosion. Finally, the author takes a look at the area's evacuation plans and concludes that "the fact that Harrisburg was not evacuated on 3/30 or before implies very strongly that a precautionary evacuation will never be ordered in a reactor accident." He feels that "when a real N disaster happens in this country, many of the victims will be evacuated directly to hospitals, where they will die."

18. "The Springtime of Our Discontent: Lessons From TMI," by Wallace E. Fisher, Sermon for 18 November, 1979, Trinity Lutheran Church, Lancaster, Pa.

The pastor reviews what he terms is his "clearest Christian thinking on that historic event" at TMI. He divides his discussion into two parts. The first part defines the cultural context in which we must make our decision about N power for electricity. The second section is devoted to the pastor's stance on N power for energy.

The major points brought up in Fisher's first section consist of: 1) distinguishing between the use of N power for destructive purposes (N armaments) and N power for other purposes, such as medicine, and 2) the increasingly technological society in which we live.

Fisher then deals with the question of "whether it is possible to have a truly representative government in a technological age." In short, the author stresses the necessity for the government to inform and consult with the citizenry on issues concerning N weapons as well as N power for energy.

A third issue presented by the pastor explains the reality of sin. Here, the author refers to TMI specifically. He comments on the "finite, ego-centered human beings" that operate the plant and explains that the human errors committed as a result of human imperfection may have more disasterous consequences "the next time."

Fisher goes on to discuss guilt, "happiness that is based on the pleasure principle," and in general the sinfulness of our society.

The part devoted to Fisher's stance on N energy begins, "I am opposed to the construction of any new plants. I am opposed to the completion of plants presently under construction." The remainder of the article discusses the pastor's arguments against N energy and his proposed alternatives.

19. "The Accident At Three Mile Island: Social Science Perspectives," D.P. Wolfe, *Social Impact Assessment*, 47/48, November/December 1979, pp. 3-8.

This is an account of the Social Science Research Council's preparation of a report designed to identify and illustrate areas of social inquiry useful to an investigation of the accident at TMI. The author considers the social perspective on the accident to be important and recognizes that the Kemeny Commission neglected to include this perspective in its scope. Therefore, a panel of consultants selected from "the Council" was directed to investigate the accident from a social standpoint.

Wolfe writes that the Commission viewed the accident at TMI as primarily a social systems failure. The Commission recorded in its report that in review of the evidence, "the fundamental problems are people related problems and not equipment problems."

Several summary statements of "the Council's" report which are divided into four parts -- organizational behavior, regulation, public participation, and conflict and consensus -- are presented at the conclusion of the article.

20. "Demythologizing Disasters," paper presented by Harry Etzkowitz, to Eastern Sociological Society, March, 1981, New York City.

This paper discusses the conflation between natural and man-made events and between discrete and long term effects in theorizing about disasters. The author states that it is this conflation that encourages the "mistaken social science analysis" that events such as TMI are accidental.

Etzkowitz then shows by defining the term "accident" that TMI was actually not an accident, but a "malfunctioning of the machinery of an industrial system." In other words, if an accident is to be described as something unforeseen, then TMI was not an accident, since the N industry studies predicted a breakdown. The author suggests that breakdowns such as the one that occurred at TMI are better viewed as inevitable products or outputs of a system of production, rather than as accidents.

Etzkowitz stresses that if TMI is to be considered an accident, "then the continued existence of the N industry is implicitly legitimated." However, if TMI is to be conceptualized as a product of an industrial system that could produce similar events, then the possibility of terminating the N energy industry would surface.

Listing of Books Dealing With Three Mile Island

In this section, a partial listing of the books that have been published since the TMI accident is provided. In addition to books specifically related to TMI, there are several books that deal with N energy in general.

1. Asimov, Issac, *Worlds Within Worlds: The Story of Nuclear Energy*, (New York: International Scholastic Book Service, 1980).

2. Calder, N., *Nuclear Nightmares*, (New York: Penquin Books, 1981).

3. Caldicott, Helen, *Nuclear Madness: What You Can Do*, (Brookline: Autumn Press, 1979).

4. Del Tredici, Robert, *The People of Three Mile Island*, (San Francisco: Sierra Club Books, 1980).

5. Freeman, L.J., *Nuclear Witnesses: Insiders Speak Out*, (New York City: W.W. Norton & Co., 1981).

6. Greenhalgh, G., *The Necessity for Nuclear Power*, (New York: Crane-Russak Company, 1981).

7. Keisling, Bill, *TMI*, (Trumansburg, NY: Crossing Press, 1980).

8. Leppzer, Robert, *Voices From TMI: The People Speak Out*, (Trumansburg, NY: Crossing Press, 1980).

9. Lillenthal, David, *Atomic Energy: A New Start*, (New York: Harper-Row Company, 1980).

10. Martin, Daniel, *Three Mile Island: Prologue or Epilogue*, (Cambridge, Mass.: Ballinger Publishing Co., 1980).

11. Miller, Jack, *A Primer on Nuclear Power*, (Millville, Minn.: Anvil Press, 1981).

12. Moss, Thomas, and Sills, David, eds., *The TMI Nuclear Accident: Lessons and Implications*, (New York: New York Academy of Science, 1981).

13. Nader, R., and Abbots, J., *The Menace of Atomic Energy*, (New York: W.W. Norton and Co., 1979).

14. Okrent, David, *Nuclear Reactor Safety: On the History of the Regulatory Process*, (Madison, Wisconsin: University of Wisconsin Press, 1981).

15. Osterhout, M., ed., *Decontamination and Decommissioning of Nuclear Facilities*, (New York: Plenum Publishing Co., 1980).

16. Pringle, P., and Spigelman, J.J., *The Nuclear Barons*, New York: Holt, Rinehart, & Winston, 1981).

17. Rolph, E.S., *Nuclear Power and the Public Safety: A Study in Regulation*, (Lexington, Mass.: Lexington Books, 1979).

18. Sills, David, *et. al.*, *Accident at Three Mile Island: The Human Dimensions*, (Boulder, Colorado: Westview Press, 1981).

19. Stephens, Mark, *TMI*, (New York: Random House, 1981).

20. Williams, M.M., *Nuclear Safety*, (Elmsford, NY: Pergamon, 1979).

21. Williams, M.M., and McCormick, N.J., eds., *Progress in Nuclear Energy: Staff Reports to the President's Commission on the Accident at TMI*, (Elmsford, NY: Pergamon, 1981).

DIRECTORY OF TMI PRO AND ANTI-NUCLEAR GROUPS

This directory is limited to groups in the TMI geographical area that have as their main purpose either advocacy or opposition in respect to TMI.

It is obvious from the listing that anti-TMI groups outnumber pro groups. The listing is as current as possible at the time of writing. One problem is that spontaneous groups often spring up around one TMI related issue e.g. dumping of water or the restart. These single purpose groups are sometimes short lived; some of them become coopted or incorporated in larger more multi-purpose groups. Pro-nuclear groups give the appearance of a more organized industrial sponsorship as opposed to anti-nuclear groups which appear to be more spontaneous and grassroots in their origins.

Some groups take the form of resource centers, transmitting information on a wide spectrum of the nuclear movement. Examples are the Atomic Industrial Forum and the Nuclear Information and Resource Service. Within the local TMI area, the TMI Public Interest Resource Center serves as an umbrella organization to smaller groups from the geographic area.

A. TMI Anti-Nuclear Groups

Environmental Coalition on Nuclear Power
433 Orlando Avenue
State College, PA 16801

Hershey Area Alliance
129 Cocoa Avenue
Hershey, PA 17033

Newberry Township TMI Steering Committee
332 Valley Road
Etters, PA 17319

Nuclear Information and Resource Service
1536 16th Street, N.W.
Washington, DC 20036

People Against Nuclear Energy
501 Vine Street
Middletown, PA 17057

Susquehanna Valley Alliance
110 Tulane Terrace
Lancaster, PA 17603

Three Mile Island Alert
315 Peffer Street
Harrisburg, PA 17102

TMI Public Interest Resource Center
1037 Maclay Street
Harrisburg, PA 17103

Union of Concerned Scientists
1384 Massachusetts Avenue
Cambridge, MA 02238

York Committee for Safe Environment
Philadelphia Street
York, PA 17315

York County Environmental Council
PO Box 1382
York, PA 17405

B. Pro-Nuclear Groups

Atomic Industrial Forum
7101 Wisconsin Avenue, N.W.
Washington, DC 20014

Friends and Family of TMI
PO Box 82
Highspire, PA 17034

GPU Nuclear Corporation
Public Information Department
Three Mile Island Nuclear Station
PO Box 480
Middletown, PA 17057

FEDERAL GOVERNMENT LIBRARY RESOURCES

Traditionally, various libraries in the U.S. have been designated as federal government depository libraries. These libraries have collections of federal government publications. It should, however, be noted that the designated library is not required to accept every government publication. Therefore, it is possible that the library will not be a depository for NRC documents.

In section A, the reader will find the names, addresses, and telephone numbers for the federal government depository libraries. This listing appeared in a March 1981 U.S. Government Printing Office publication on government depository libraries.

In section B, the reader will be directed to the name and addresses of libraries serving as depositories for NRC documents. These libraries were established by the NRC after the accident at TMI. Each library specializes in documents associated with the nuclear power plant in their area. Most documents originated by NRC, or submitted to it for consideration including letters, reports, and testimony are placed in these designated libraries. This list appeared in the 1980 Annual Report of the USNRC. For a current list, the reader can contact the Local Public Document Room Branch, Division of Rules and Records, USNRC, Washington, D.C. 20555.

A. Government Depository Libraries
Revised March 1981

Auburn University at Montgomery Library
REGIONAL DEPOSITORY
Documents Department
Montgomery, AL 36109
(1971)
(205) 279-9110, Ext. 253

Department of Library Archives, and Public Records
REGIONAL DEPOSITORY
Third Floor State Capitol
Phoenix, AZ 85007
(602) 255-4142

California State Library
REGIONAL DEPOSITORY
Documents Section
PO Box 2037
Sacramento, CA 95809
(1895)
(916) 322-4572

University of Colorado Library
REGIONAL DEPOSITORY
Government Documents Division
Boulder, CO 80309
(1879)
(303) 492-8834

Connecticut State Library
REGIONAL DEPOSITORY
231 Capitol Avenue
Hartford, CT 06115
(203) 566-4971

Library of Congress
Serial and Governments' Publications
Documents Unit, Room 1040 H. TJB
Washington, D.C. 20540
(1977)
(202) 426-5251

University of Florida Libraries
REGIONAL DEPOSITORY
Documents Department
Gainesville, FL 32601
(1907)
(904) 392-0367

University of Georgia Libraries
REGIONAL DEPOSITORY
Documents Section
Athens, Georgia 30602
(1907)
(404) 542-8951

University of Hawaii Library
REGIONAL DEPOSITORY
Government Documents Collection
2550 The Mall
Honolulu, Hawaii 96822
(1907)
(808) 948-8230

Illinois State Library
REGIONAL DEPOSITORY
Government Documents
Centennial Building
Springfield, Illinois 62706
(217) 782-5185

Indiana State Library
REGIONAL DEPOSITORY
Serials Section
140 North Senate Avenue
Indianapolis, Indiana 46204
(317) 232-3686

University of Iowa Libraries
REGIONAL DEPOSITORY
Government Documents Department
Iowa City, Iowa 52242
(1884)
(319) 353-3318

University of Kansas
REGIONAL DEPOSITORY
Watson Library
Documents Collection
Lawrence, Kansas 66045
(1869)
(913) 864-4662

University of Kentucky Libraries
REGIONAL DEPOSITORY
Government Publications Department
Lexington, Kentucky 40506
(1907)
(606) 257-2639

Louisiana State University Library
REGIONAL DEPOSITORY
Government Documents Department
Baton Rouge, Louisiana 70803
(1907)
(504) 388-2570

University of Maine
REGIONAL DEPOSITORY
Raymond H. Fogler Library
Tri-State Regional Documents Depository
Orono, Maine 04469
(1907)
(207) 581-7178

University of Maryland
REGIONAL DEPOSITORY
McKeldin Library
Documents Division
College Park, Maryland 20742
(1925)
(301) 454-3034

Boston Public Library
REGIONAL DEPOSITORY
Documents Department
Boston, Massachusetts 02117
(1859)
(617) 536-5400, Ext. 295

Detroit Public Library
REGIONAL DEPOSITORY
5201 Woodward Avenue
Detroit, Michigan 48202
(1868)
(313) 833-1409

University of Minnesota
REGIONAL DEPOSITORY
Wilson Library
Documents Division
Minneapolis, Minnesota 55455
(1907)
(612) 373-7813

University of Mississippi Library
REGIONAL DEPOSITORY
Documents Department
University, Mississippi 38677
(1833)
(601) 232-7091, Ext. 7

Central Missouri State University
Ward Edwards Library
Government Documents
Warrensburg, Missouri 64093
(1914)
(816) 429-4149

University of Montana Library
REGIONAL DEPOSITORY
Documents Department
Missoula, Montana 59812
(1909)
(406) 243-6700

Nebraska Publications Clearinghouse
REGIONAL DEPOSITORY, in cooperation with
University of Nebraska at Lincoln
D.L. Love Memorial Library
1420 P Street
Lincoln, Nebraska 68508
(1972)
(308) 471-2045

University of Nevada Library
REGIONAL DEPOSITORY
Government Publications Department
Reno, Nevada 89557
(1907)
(702) 784-6579

Newark Public Library
REGIONAL DEPOSITORY
5 Washington Street
Newark, New Jersey 07107
(1906)
(201) 733-7812

University of New Mexico
REGIONAL DEPOSITORY
Zimmerman Library
Albuquerque, New Mexico 87131
(1896)
(505) 277-5441

New York State Library
REGIONAL DEPOSITORY
Documents Control
6th Floor, Cultural Education Center
Empire State Plaza
Albany, New York 12230
(518) 474-5563

University of North Carolina at Chapel Hill
Library
REGIONAL DEPOSITORY
BA/SS Documents Division
Chapel Hill, North Carolina 27515
(1884)
(919) 933-1151

North Dakota State University Library
REGIONAL DEPOSITORY, in cooperation with
University of North Dakota Chester Fritz
Library
Government Documents Department
Fargo, North Dakota 58105
(1907)
(701) 237-8886

Oklahoma Department of Libraries
REGIONAL DEPOSITORY
Government Documents
200 NE 18th Street
Oklahoma City, Oklahoma 73105
(1893)
(405) 521-2502

State Library of Pennsylvania
REGIONAL DEPOSITORY
Government Publications Section
Box 1601
Harrisburg, PA 17126
(717) 787-3752

Texas State Library
REGIONAL DEPOSITORY
Documents Section
PO Box 12927
Capitol Station
Austin, Texas 78711
(512) 475-2996

Utah State University
REGIONAL DEPOSITORY
Merrill Library and Learning
Resources Center UMC-30
Logan, Utah 84322
(1907)
(801) 750-2682

University of Virginia
REGIONAL DEPOSITORY
Alderman Library
Public Documents
Charlottesville, Virginia 22903
(1910)
(804) 924-3133

Washington State Library
REGIONAL DEPOSITORY
Documents Section
Olympia, Washington 98504
(206) 753-6525

West Virginia University Library
REGIONAL DEPOSITORY
Documents Department
Morgantown, W. Virginia 26506
(1907)
(304) 293-5440

State Historical Society Library
REGIONAL DEPOSITORY, in cooperation with
University of Wisconsin-Madison
Documents Serials Section
Memorial Library
816 State Street
Madison, Wisconsin 53706
(1870)
(608) 262-4347

Wyoming State Library
REGIONAL DEPOSITORY
Supreme Court and Library Building
Cheyenne, Wyoming 82001
(307) 777-7281, Ext. 26

B. NRC's Public Document Rooms

Alabama

Mrs. Maude S. Miller
Athens Public Library
South and Forrest
Athens, AL 35611
Browns Ferry N Plant

Mr. Wayne Love
G.S. Houston Mem Library
212 W. Burdeshaw St.
Dothan, AL 36303
Farley N Plant

Mrs. Joanne Wyatt
Clanton Public Library
100 First St.
Clanton, AL 35045
Barton N Plant

Mrs. Peggy McCutchen
Scottsboro Public Library
1002 S. Broad St.
Scottsboro, AL 35768
Bellefonte N Plant

Ms. Mary Ann Lovell
Prattville Public Library
220 Doster Road
Prattville, AL 36067
AL N Fuel Fabrication

Arizona

Mrs. Mary Carlson
Phoenix Public Library
Science & Industry Sec.
12 E. McDowell Rd.
Phoenix, AZ 85004
Palo Verde N Plant

Arkansas

Mr. Vaughn
AR Polytechnic College
Russellville, AR 72801
AR N One

California

Mrs. Alice Rosenberger
Palo Verde Valley Dist. Library
125 West Chanslorway
Blythe, CA 92255
Sundesert N Plant

Mr. William B. Rohan
San Diego County Law Library
1105 Front St.
San Diego, CA 92101
Sundesert N Plant

Mrs. Eileen Danforth
Mission Viejo Branch Library
24851 Chrisanta Dr.
Mission Viejo, CA 92676
San Onofre N Plant

Mr. Chi Su Kim
Documents & Maps Dept.
CA Polytechnic State University Library
San Luis Obispo, CA 93407
Diablo Canyon N Plant

Mrs. Judy Klapprott
Humboldt County Library
636 F Street
Eureka, CA 95501
Humboldt Bay N Plant

Mrs. Gabrielle Holmes
Business & Municipal Dept.
Sacramento City-County Library
828 I St.
Sacramento, CA 95814
Rancho Seco N Plant

Stanislaus County Free Library
1500 I St.
Modesto, CA 95345
Stanislaus N Plant

NRC, Region V
Suite 202
1990 N. California Blvd.
Walnut Creek, CA 94596
GETR Vallecitos

W. Los Angeles Regional Library
11360 Santa Monica Blvd.
Los Angeles, CA 94596
UCLA Research Reactor

Colorado

Miss Ester Fromm
Greeley Public Library
City Complex Building
Greeley, CO 80631
Fort St. Vrain N Plant

Mrs. Robin Satterwhit
Government Documents
Auraria Library
Univ. of CO at Denver
Lawrence & 11th
Denver, CO 80204
Atlis Corp. Uranium Mill

Connecticut

Mr. Vincent Juliano
Waterford Public Library
Rope Ferry Rd.-Route 156
Waterford, CT 06385
Millstone N Plant

Mrs. Phyllis Nathanson
Russell Library
119 Broad St.
Middletown, CT 06457
Haddam Neck N Plant

Delaware

Mrs. Yvonne Puffer
Newark Free Library
750 E. Delaware Ave.
Newark, DE 19711
Summit N Plant

Florida

Ms. Sally Litton
Jacksonville Public Library
122 N. Ocean St.
Jacksonville, FL 32204
Offshore Power Systems Manufacturing Facility

Mrs. R. Scott
Indian River Community College Library
3209 Virginia Ave.
Ft. Pierve, FL 33450
St. Lucie N Plant

Mrs. Bonsall
Crystal River Public Library
668 NW First
Crystal River, FL 32629
Crystal River N Plant

Mrs. Rene Daily
Environmental & Urban Affairs Library
FL International Univ.
Miami, FL 33199
Turkey Point N Plant

Ms. Renee Pierce
Lily Lawrence Bow Library
212 NW First Ave.
Holmstead, FL 33030
Turkey Point N Plant
(Emergency Plan Only)

Georgia

Mrs. J.W. Borom
Burke County Library
Fourth Street
Waynesboro, GA 30830
Vogtle N Plant

Ms. Annette Osborne
Appling County Public Library
301 City Hall Dr.
Baxley, GA 31513
Hatch N Plant

Illinois

Mr. Ed Anderson
IL Valley Community College
Rural Route #1
Oglesby, IL 16348
LaSalle N Plant

Mrs. Pam Wilson
Morris Public Library
604 Liberty Street
Morris, IL 60451
Dresden N Plant-Midwest Fuel Recovery Plant

Mrs. Marie Hoschied
Moline Public Library
504 17th Street
Moline, IL 61255
Quad Cities N Plant

Ms. Jo Ann Ellingson
Zion-Benton Public Library
2600 Emmaus Ave.
Zion, IL 60099
Zion N Plant

Mrs. M. Evans
Vespasian Warner Public Library
120 W. Johnson St.
Clinton, IL 61727
Clinton N Plant

Ms. Kay Sauer
West Chicago Public Library
322 E. Washington St.
West Chicago, IL 60185
Rare Earth Facility

Mrs. Penny O'Roarke
Byron Public Library
Third & Washington Sts.
Byron, IL 61010
Byron N Plant

Mr. Thomas Carter
Wilmington Twp. Public Library
201 S. Kankakee St.
Wilmington, IL 60481
Braidwood N Plant

Savanna Twp. Public Library
326 Third St.
Savanna, IL 61074
Carroll N Plant

Mr. Richard Gray
Rockford Public Library
215 N. Wyman St.
Rockford, IL 61103
Byron N Plant

Indiana

Ms. Michele Stipanovich
W. Chester Twp. Public Library
125 S. Second St.
Chestertown, IN 46304
Bailly N Plant

Ms. Carol Cowles
Madison-Jefferson County Public Library
420 W. Main St.
Madison, IN 47250
Marble Hill N Plant

Iowa

Ms. Linda Hanley
Reference Service
Cedar Rapids Public Library
428 Third Ave., SE
Cedar Rapids, IA 52401
Duane Arnold N Plant

Kansas

Mr. Jack Scott
Coffey County Courthouse
Burlington, KS 66839
Wolf Creek N Plant

Kentucky

Mr. Clarence R. Graham
Louisville Free Public Library
4th & York Sts.
Louisville, KY 40203
Marble Hill N Plant

Ms. Beverly Bury
Campbell County Public Library
Alexandria Branch
400 W. Main St.
Alexandria, KY 41001
Zimmer N Plant

Louisiana

Mr. Ken Owen
Univ. of New Orleans Library
Louisiana Collection, Lakefront
New Orleans, LA 70122
Waterford N Plant

Mrs. Freeda Fisher
Audubon Library-West Feliciana Branch
Ferdinand Street
St. Francisville, LA 70775

Mr. Jimmie H. Hoover
Government Documents Dept.
Louisiana State Univ.
Baton Rouge, LA 70803
River Bend N Plant

Maine

Mrs. Barbara Shelton
Wiscasset Public Library
High Street
Wiscasset, ME 04578
ME Yankee N Plant

Maryland

Mrs. Elizabeth Hart
Charles County Library
Garrett & Charles Sts.
La Plata, MD 20646
Douglas Point N Plant

Mrs. Marie Barrett
Calvert County Library
Prince Frederick, MD 20678
Calvert Cliffs N Plant

Ms. Margaret Jacobs
Enoch Pratt Free Library
Business, Sciences & Technology Dept.
Central Library
400 Cathedral Street
Baltimore, MD 21201
TMI-1 Suspension Proceeding (Transcripts Only)

Massachusetts

Mrs. Margaret Howland
Greenfield Community College
One College Drive
Greenfield, MA 01301
Yankee Rowe N Plant

Ms. Ruth Chamberlain
Plymouth Public Library
North Street
Plymouth, MA 02360
Pilgrim N Plant

The Carnegie Library
Avenue A
Turner Falls, MA 01376
Montague N Plant

Michigan

Mrs. Diana Shamp
Reference Department
Kalamazoo Public Library
315 S. Rose St.
Kalamazoo, MI 49006
Palisades N Plant

Mrs. Katherine Thomson
St. Clair County Library
210 McMorran Blvd.
Port Huron, MI 48060
Greenwood N Plant

Mrs. M.B. Wallick
Charlevoix Public
Library
107 Clinton St.
Charlevoix, MI 49720
Big Rock Point

Mrs. Averill Packard
Grace Dow Mem Library
1710 W. St. Andrews Rd.
Midland, MI 48640
Midland N Plant

Ms. Ann Stobbe
Maude Preston Palenske
Mem Library
500 Market St.
St. Joseph, MI 49085
D.C. Cook N Plant

Mrs. Sarah Peth
Reference Department
Monroe County Library
System
3700 S. Custer Rd.
Monroe, MI 48161
Fermi N Plant

Minnesota

Mrs. Copeland
Environmental Conservation Library
Minneapolis Public
Library
300 Nicollet Mall
Minneapolis, MN 55401
Monticello N Plant
Prairie Island N Plant

Missouri

Mrs. Ladonna Justice
Fulton City Library
709 Market St.
Fulton, MO 65251
Callaway N Plant

Mrs. Ranata Rotkowicz
Olin Library of Washington University
Skinker & Lindell Blvd.
St. Louis, MO 63130
Callaway N Plant

Mississippi

Mr. William McMullin
Corinth Public Library
1023 Fillmore St.
Corinth, MS 38834
Yellow Creek N Plant

Nebraska

Mr. Frank Gibson
W. Dale Clark Library
215 S. 15th St.
Omaha, NE 68102
Ft. Calhoun N Plant

Mrs. Loy Mowery
Auburn Public Library
118 15th Street
Auburn, NE 68305
Cooper N Plant

New Hampshire

Miss Pamela Gjettum
Exeter Public Library
Front Street
Exeter, NH 03883
Seabrook N Plant

New Jersey

Mrs. Ginny Vail
Stockton State College Library
Pomona, NJ 08240
  Offshore Power Systems Manufacturing Facility

Miss Elizabeth Fogg
Salem Free Public Library
112 W. Broadway
Salem, NJ 08097
  Salem N Plant
  Hope Creek N Plant

Mrs. Dolores Waddill
Ocean County Library
Brick Twp. Branch
401 Chambers Bridge Rd.
Brick, NJ 08723
  Oyster Creek N Plant
  Forked River N Plant

New Mexico

Ms. Sandra Coleman
General Library, Ref. Dept.
Univ. of New Mexico
Albuquerque, NM 87131
  Waste Isolation Pilot Plant

Ms. Ingrid Vollnhofer
New Mexico State Library
Box 1629
Sante Fe, NM 87503
  Waste Isolation Pilot Plant

New York

Documents Librarian
Penfield Library
State Univ. College at Oswego
Oswego, NY 13126
  Nine Mile Point N Plant
  FitzPatrick N Plant
  New Haven N Plant

Mrs. June Rogoff
Rochester Public Library
Business & Social Science Division
115 South Avenue
Rochester, NY 14604
  Ginna N Plant

Mr. Oliver Swift
White Plains Public Library
100 Martine Avenue
White Plains, NY 10601
  Indian Point N Plant

Mr. Peter Allison
New York University
70 Washington Sq. S.
New York, NY 10012
  (1979 and Later Material)

Kathy McGowan
Shoreham-Wading River Public Library
Route 25A
Shoreham, NY 11786
  Shoreham N Plant

Mrs. E. Overton
Riverhead Free Library
330 Court Street
Riverhead, NY 11901
  Jamesport N Plant

Mr. Stanley Zukowzki
Buffalo & Erie County Public Library
Lafayette Square
Buffalo, NY 14203
  NFS Fuel Reprocessing Plant and UF_6 Facility

Ms. Marsha Russell
Town of Concord Public Library
23 N. Buffalo St.
Springville, NY 14141
NFS Fuel Reprocessing Plant and UF_6 Facility

Mr. Sol Becker
Public Health Library
NYC Dept. of Health
125 Worth Street
New York, NY 10012
Columbia Univ. Research Reactor

Mrs. Dorothy Augustine
Catskill Public Library
One Franklin St.
Catskill, NY 12414
Greene County N Plant

Mr. Harold Ettelt
Columbia-Greene Community College
PO Box 100
Hudson, NY 12534
Greene County N Plant (Transcripts Only)

North Carolina

Ms. Dawn Hubbs
Atkins Library
Univ. of NC-Charlotte
UNCC Station, NC 28223
McGuire N Plant

Mr. Roy Dicks
Wake County Public Library
104 Feyetteville St.
Raleigh, NC 27601
Shearon Harris N Plant

Mr. David G. Ferguson
Davie County Public Library
416 N. Main St.
PO Box 158
Mocksville, NC 27028
Perkins N Plant

Southport-Brunswick County Library
109 W. Moore St.
Southport, NC 28461
Brunswick N Plant

Mrs. Ann Laliotes
Franklin County Library
1026 Justice Street
Louisburg, NC 27549
Gulf Youngsville Fuel Fabrication Facility

Ohio

Mrs. Betty Waltman
Perry Public Library
3753 Main Street
Perry, OH 44081
Perry N Plant

Mrs. Mary Mackzum
Clermont County Library
Third and Broadway Sts.
Batavia, OH 45103
Zimmer N Plant

Mr. Donald Fought
Ida Rupp Public Library
310 Madison Street
Port Clinton, OH 43452
Davis-Besse N Plant

Oklahoma

Mr. Craig Buthod
Tulsa City-County Library
400 Civic Center
Tulsa, OK 74102
Black Fox N Plant

Mrs. O.J. Grosclaude
Sallisaw City Library
111 North Elm
Sallisaw, OK 74955
Sequoyah UF_6 Facility

Mrs. Carol Robinson
Guthrie Public Library
201 North Division
Guthrie, OK 73044
Cimarron Pu Fabrication Plant and Uranium Fuel Facility

Oregon

Miss Carol VonDerAhe
City Hall, Records Office
Arlington, OR 97812
Pebble Springs N Plant

Mr. Jim Takita
Multnomah County Library
Social Science Dept.
801 SW 10th Ave.
Portland, OR 97205
Trojan N Plant

Pennsylvania

Mrs. Gail Frew
Reference Department
Osterhout Free Library
71 S. Franklin St.
Wilkes-Barre, PA 18701
Susquehanna N Plant

PA State University
Central Pattee Library
Room 207
University Park, PA 16802
Susquehanna N Plant
(Transcripts Only)

Ms. Connie Webster
E. Shore Area Branch Library
4501 Ethel Street
Harrisburg, PA 17109
TMI N Plant
(Transcripts Only)

Mrs. Clifford Crowers
Free Library of Phila.
Gov. Publications Dept.
19th & Vine
Phila., PA 19103
TMI N Plant
(Transcripts Only)

Ms. Elizabeth Harvey
Schlow Mem Library
100 E. Beaver Ave.
State College, PA 16801
TMI N Plant
(Transcripts Only)

Mr. John Geschwindt
Gov. Publications Section
State Library of PA
Education Building
Commonwealth & Walnut Sts.
Harrisburg, PA 17126
Peach Bottom N Plant
TMI N Plant
Fulton N Plant

Mrs. Gordon Bauerie
Pottstown Public Library
500 High St.
Pottstown, PA 19464
Limerick N Plant

Apollo Mem Library
219 N. Pennsylvania Ave.
Apollo, PA 15613
Apollo UF_6 and Pu Facilities

Mrs. Catherine Brosky
Carnegie Library of Pittsburgh
4400 Forbes Ave.
Pittsburgh, PA 15213
Cheswick Fuel Dev. Laboratories

Mrs. Mary Columbo
B.F. Jones Mem Library
663 Franklin Ave.
Aliquippa, PA 15001
Beaver Valley N Plant
Shippingport Light Water Breeder Reactor

Puerto Rico

Mrs. Rosario Cabrera
Public Library, City Hall
Jose de Diego Ave.
PO Box 1086
Arecibo, PR 00612
North Coast N Plant

Mrs. Amalia Ruiz De Porras
Etien Totti Public Library
College of Engineers, Architects & Surveyors
Urb Roosevelt Dev.
Hato Rey, PR 00918
North Coast N Plant

Rhode Island

Mrs. Ann Crawford
Cross Mill Public Library
Old Post Road
Charlestown, RI 02831
Wood River Junction

Mr. Thomas Reynolds
Univ. of Rhode Island
University Library
Gov. Publications Office
Kingston, RI 02881
Wood River Junction

South Carolina

Ms. Mary Mallaney
York County Library
325 S. Oakland Avenue
Rock Hill, SC 29730
Catawba N Plant

Mr. Ed Kilroy
Oconee County Library
401 W. Southbroad
Walhalla, SC 29691
Oconee N Plant

Mrs. Peggy Cover
Clemson Univ. Library
Science, Technology & Agricultural Services
Clemson, SC 29631
Oconee N Plant
(Limited Documentation)

Reference Department
Richland County Public Library
1400 Sumter Street
Columbia, SC 29201
Summer N Plant

Mrs. Allene Reep
Hartsville Mem Library
Home & Fifth Aves.
Hartsville, SC 29550
H.B. Robinson N Plant

Mr. David Eden
Cherokee County Library
300 E. Rutledge Ave.
Gaffney, SC 29340
Cherokee N Plant

Mr. T.E. Richardson
County Office Building
Room 105
PO Box 443
Barnwell, SC 29812
Barnwell Fuel Plant
UF_6 Facility
Barnwell Fuel Storage Station

Mr. Carl Stone
Anderson County Library
202 E. Greenville St.
Anderson, SC 29621
Recycle Fuel Plant

Mrs. Ellen Jenkins
Barnwell County Library
Hagood Avenue
Barnwell, SC 29812
Chem-N Plant

Tennessee

Miss Kendall J. Cram
TN State Library & Archives
403 Seventh Ave., N.
Nashville, TN 37219
Hartsville N Plant

Ms. Dorothy Dismuke
Oak Ridge Public Library
Civic Center
Oak Ridge, TN 37830
Clinch River Breeder Plant
Exxon N Fuel Recovery Center

Mrs. Patricia Rugg
Lawson McGhee Public Library
500 W. Church St.
Knoxville, TN 37902
Clinch River Breeder Plant
Exxon N Fuel Recovery Center
Fuel Fabrication Facility

Mr. Wally Keasler
Chattanooga-Hamilton County Bicentennial Library
1001 Broad St.
Chattanooga, TN 37402
Sequoyah N Plant
Watts Bar N Plant

Mr. T. Cal Hendrix
Kingsport Public Library
Broad & New Sts.
Kingsport, TN 37660
Phipps Bend N Plant

Mr. H.E. Zittel
Oak Ridge Nat. Laboratory
PO Box X
Oak Ridge, TN 37830
Tyrone N Plant
(Transcripts Only)

Texas

Mr. John Hudson
Univ. of Texas at Arlington
Arlington, TX 76019
Comanche Peak N Plant
(Limited Documentation)

Ms. May Schmidt
Austin-Travis County Collection
Austin Public Library
810 Guadalupe St.
PO Box 2287
Austin, TX 78768
South Texas N Plant

Matagorda County Courthouse
Matagorda County Law Library
PO Box 487
Bay City, TX 77414
South Texas N Plant

Mr. James Sosa
San Antonio Public Library
Business, Science & Technology Dept.
203 S. St. Mary St.
San Antonio, TX 78205
South Texas N Plant
(Inspection Reports Only)

Mrs. Tim Whitworth
Somervell County Public Library
On The Square
PO Box 1417
Glen Rose, TX 76043
Comanche Peak N Plant

Newton County Library
PO Box 657
Newton, TX 77034
Blue Hills N Plant

Mrs. Kroesche
Sealy Public Library
201 Atchison Street
Sealy, TX 77474
Allens Creek N Plant

Vermont

Mrs. June Bryant
Brooks Mem Library
224 Main St.
Brattleboro, VT 05301
Vermont Yankee N Plant

Virginia

Ms. Sandra Peterson
Swem Library
College of William & Mary
Williamsburg, VA 23185
Surry N Plant

Mr. Edward Kube
Board of Supervisors
Louisa County Courthouse
PO Box 27
Louisa, VA 23093
North Anna N Plant

Mr. Gregory Johnson
Alderman Library
Manuscripts Department
University of Virginia
Charlottesville, VA 22901
North Anna N Plant

Washington

Ms. D.E. Roberts
Richland Public Library
Swift & Northgate Sts.
Richland, WA 99352
WPPSS 1, 2 & 4 N Plants
Skagit N Plant
Exxon Fuel Plants

Mrs. Mary Ann Schafer
W.H. Abel Mem Library
125 Main Street South
Montesano, WA 98563
WPPSS 3 & 5 N Plants

Wisconsin

Mrs. Jane Radloff
LaCrosse Public Library
800 Main St.
LaCrosse, WI 54601
LaCrosse BWR N Plant

Elsie Heitkemper
Joseph Mann Library
1516 Sixteenth Street
Two Rivers, WI 54241
Point Beach N Plant

Mr. Arthur M. Fish
Document Dept, Library
Univ. of Wisconsin
Stevens Point
Stevens Point, WI 54481
Point Beach N Plant
(Limited Documentation)
Wood N Plant

Ms. Sue Grossheuch
Kewaunee Public Library
822 Juneau St.
Kewaunee, WI 54216
Kewaunee N Plant

Mr. John Jax
Univ. of Wisconsin
Stout Library
Menomonie, WI 54741
Tyrone N Plant

Mr. Robert Fetvedt
University Library
Univ. of Wisconsin-Eau Claire
Park & Garfield Aves.
Eau Claire, WI 54710
Tyrone N Plant
(Transcripts Only)

Mrs. Robert Goodrich
Durand Free Library
315 Second Ave., W.
Durand, WI 54736
Tyrone N Plant
(Transcripts Only)

Wyoming

Mrs. Carroll Highfill
Converse County Library
Douglas, WY 82633
Highland Uranium Mill

Mrs. Bess Sheller
Carbon County Public Library
Courthouse
Rawlins, WY 82301
Shirley Basin Uranium Mill

GLOSSARY

The annotations of the media coverage and government documents dealing with TMI use many terms that are specific to the N industry. In order to aid the reader, we are providing the glossary included in the *Report of the Governor's Commission on Three Mile Island*. This Commission was chaired by Lt. Gov. William Scranton, III. The Report is dated 2/26/80.

ASLB-A board appointed by the NRC to conduct the licensing proceedings for new N power plants, as the need arises.

Auxiliary building-A structure housing a variety of equipment and large tanks necessary for the operation of the reactor.

Background radiation-Radiation arising from natural radioactive materials always present in the environment, including solar and cosmic radiation and radioactive elements in the upper atmosphere, the ground, building materials, and the human body.

Beta particles-High energy electrons; a form of ionizing radiation that normally is stopped by the skin, or a very thin sheet of metal.

Central Penn Multi-List, Inc.-A listing of all property for sale by member realtors in the greater Harrisburg area.

Cesium 134-Radioactive form of cesium, with a half-life of two years.

Cesium 137-A radioactive form of cesium, with a half-life of 30 years. Emits both gamma and beta radiation.

Class-action suit-A legal action undertaken by one or more plaintiffs on behalf of themselves or others having an identical interest in the alleged wrong.

Congenital/neonatal hypothyroidism-A condition present at birth or within the first month after birth in which there is deficient activity of the thyroid gland, resulting in a lowered metabolic rate and general loss of vigor.

Containment building-The structure housing the N reactor; intended to contain radioactive solids, gases, and water that might be released from the reactor vessel in an accident.

Core-The central part of a nuclear reactor that contains the fuel and produces the heat.

Debenture-A certificate or voucher acknowledging a debt.

Disaster Operations Plan-A written response plan for all types of emergencies and disasters occurring within the Commonwealth. Prepared and implemented by the PEMA.

Duty Officer-A person who provides responsible coverage for the designated Commonwealth agency during nonworking hours.

Econometric-Application of statistical methods to the study of economic data and problems.

Endocrinologist-A scientist specializing in the study of the endocrine glands.

Environmental assessment report-An evaluation of the environmental impact of the stated activity.

Epidemiologist-A scientist specializing in study of the incidence, distribution, and control of disease in a population.

Federal Disaster Relief Act-A special Congressional act providing federal assistance to state and local governments during emergencies and major disasters.

Fission-The splitting apart of a heavy atomic nucleus into two or more parts when a neutron strikes the nucleus. The splitting releases a large amount of energy.

Fuel handling building-One of the adjacent structures to the containment building where uranium fuel rods are stored.

Gamma rays-High energy electromatic radiation; a form of ionizing radiation of higher energy than X-rays that penetrates very deep into body tissues.

General emergency-Declared by the utility when an incident at a N power plant poses a potentially serious threat of radiation releases that could affect the general public.

Genetic diseases or defects-Health defects inherited by a child from the mother and/or father.

Half-life-The time required for half of a given radioactive substance to decay. The radioactivity of an isotope with a half-life of five days would be reduced by one-half in a five day period. After the second five day period, the radioactivity would be one-fourth of the original, and so on.

Health physics-The practice of protecting humans and their environment from the possible hazards of radiation.

Hydrogen bubble-A volume of hydrogen gas in the top of the reactor vessel.

Iodine 131-A radioactive form of iodine, with a half-life of 8.1 days, that can be absorbed by the human thyroid if inhaled or ingested and cause noncancerous or cancerous growth.

Ion-An atom or group of atoms that carries a positive or negative charge.

Ion Exchange-A chemical reaction involving the exchange of ions present in a solid with ions of like charge present in a surrounding solution. Used in the Epicore II system for removal of radioactive isotopes from the water.

Intervenor-One who intervenes as a third party in a legal proceeding.

Krypton 85-A radioactive noble gas, with a half-life of 10.7 years, that is not absorbed by body tissues and is soon eliminated by the body if inhaled or ingested.

Loss-of-coolant accident-An accident involving a broken pipe, stuck-open valve, or other leak in the reactor coolant system that results in a loss of the water cooling the reactor core.

Low population zone-A NRC term to define the area around the reactor with low population density. This is the area for which evacuation had to be planned under NRC rules and regulations.

Middle Atlantic Federal Regional Council-A coordinating council for a group of federal domestic agencies.

Millirem-One-thousandth of a rem; see *rem*.

Negative pressure-Less than the pressure of the atmosphere.

Person-rem-The sum of the individual doses received by each member of a certain group or population. It is used to estimate the incremental number of health effects cases which a radiation exposure might produce in the given population. It is not used to determine which individuals in the population might be affected or in dealing with individual medical care needs.

Plume-Radioactive material released to the atmosphere from a stack or point source which dissipates with distance depending upon wind speed and other atmospheric conditions. Its form is similar to smoke released from a smoke stack.

Potassium iodide-A chemical that readily enters the thyroid gland when ingested. If taken in sufficient quantity prior to exposure to radioactive iodine, it can prevent the thyroid from absorbing any of the potentially harmful radioactive iodine 131.

Primary system-The system containing water that cools the reactor core and carries away heat. Also called the reactor coolant system.

Radiation Management Corporation-An independent company which maintains dosimetry stations around the TMI facility as a quality check of the utility's environmental surveillance program.

Radiation survey probe-A portable radiation detection device.

Reactor head-Removable top on the reactor vessel.

Reactor vessel-The steel tank containing the reactor core.

Rem-A standard unit of radiation dose. Frequently radiation dose is measured in millirems for low-level radiation; 1,000 millirems equal one rem.

Resins-Chemical compounds which selectively attract other elements and compounds. Used in the Epicor II system to attract radioactive isotopes.

Site emergency-Declared by the utility when an incident at a nuclear power plant threatens the uncontrolled release of radioactivity into the immediate area of the plant.

State Tax Equalization Board-A Commonwealth agency whose main function is to determine annually the aggregate market value of real property in the Commonwealth.

Strontium 90-A radioactive form of strontium, with a half-life of 28 years. Emits only gamma radiation.

Thermoluminescent dosimeter (TLD)-A device to measure environmental radiation.

Wet-chemistry and radiation counting room facility-Radioisotope analysis center where radiation detection equipment is located. Would contain gamma ray analyzer and equipment for chemical separation of radioisotopes for identification purposes.

Whole body scan-A detailed examination of the human body for the presence or localization of radioactive material.

Xenon 133-A radioactive noble gas with a half-life of 5.3 days that is not absorbed by the body tissues and is soon eliminated by the body if inhaled or ingested. Xenon 133 was the principle radioactive isotope released to the environment during the TMI accident.

NAME INDEX

SUBJECT INDEX